The Art
of Innovation

LESSONS IN CREATIVITY FROM IDEO,
AMERICA'S LEADING DESIGN FIRM

Tom Kelley

with Jonathan Littman

P

PROFILE BOOKS

This paperback edition published in Great Britain in 2016 by
PROFILE BOOKS LTD
3 Holford Yard
Bevin Way
London WC1X 9HD
www.profilebooks.com

Published in Great Britain by HarperCollinsBusiness, 2001
Published in the USA by Doubleday Broadway Publishing, 2001

13

Printed and bound in India by
Manipal Technologies Limited, Manipal

A CIP catalogue record for this book is available from the British Library.

ISBN 978 1 78125 614 5

MIX
Paper from
responsible sources
FSC™ C043100

Preface to the new edition

A lot has happened since *The Art of Innovation* first hit the shelves fifteen years ago, and one of the things that has changed along the way is innovation itself. Back then, creativity and innovation seemed to live primarily on the fringes of the enterprise. Designers felt like they were relegated to the "kids table" of the commercial world, while the serious business was taking place in conference rooms and boardrooms down the hall. In the ensuing years however, as global competition heated up and companies felt the need for continuous renewal, suddenly creativity was needed every day—and in every part of the organization. In other words, now creativity *is* serious business.

Another change is that most of the ideas and methodologies described in this book have now been gathered under the umbrella of "Design Thinking". By definition, design thinking involves applying the creative tools and mindset that designers have used for decades to new challenges going well beyond what has traditionally been thought of as design. Three of the key ingredients in the recipe for design thinking are empathy, experimentation, and storytelling, and you will find them all here. "Beginning with an eye" in Chapter 3 is ultimately about sharpening your empathic abilities. The prototyping in Chapter 6 is about treating life as an experiment, whether you are building a physical model or just moving your idea one step further towards successful implementation. And Chapter 10's suggestions about creating experiences are tied to reinforcing powerful stories about a brand.

While many of the examples in this book involve physical products, designers at IDEO and elsewhere have since expanded the use of these tools in an increasingly broad range of situations. For example, we've used these methodologies

on challenges ranging from envisioning the future of the kitchen for IKEA to building an innovative new school system in Peru to creating a digital strategy for the Royal Academy of Art.

Although I wrote *The Art of Innovation* primarily for business groups large and small, a delightful unintended consequence of the book was the huge number of emails I've received from *individuals* who, after reading the book, found a way to apply its ideas to their lives. Most of those notes came in more than a decade ago, but I can still remember many of them. The longest was from a Baptist minister in the American Southwest, who filled pages telling me how the book had inspired him in his work. The shortest was a four-line note from a 39-year-old Brazilian designer who started with "Sorry my English" and ended by explaining that "… the book lights my fire again." That short note—from a stranger I will never meet—stayed taped up on the wall of my office for almost a decade.

If life should be full of experiments, then this book was one of mine. It filled eighteen months of nights and weekends, while I tried my best to simultaneously fulfill my "day job" responsibilities at IDEO. As a first-time writer, I was never certain that the manuscript would be accepted at all, let alone translated into a dozen languages. So it is a great pleasure to know that Profile Books is publishing this new edition. And my sincere hope is that, somewhere in the English-speaking world, a reader encountering this latest edition will feel that it *lights their fire again*.

Tom Kelley, 2015

To my brother **David,**

who has been a roommate,

mentor, partner, boss,

and best friend.

Without him,

this book

would not

exist.

If it takes a village to raise a child, then creating a book seems to require at least a small metropolis. Like many of the innovation programs described inside, the book was very much a *team* effort. Although Jonathan and I get to put our names on the front cover, literally dozens of people have contributed to the final product. I won't try to name them all—like some Oscar Awards speech gone haywire—but several people made such significant contributions that I feel compelled to single them out.

First, literary agent Richard Abate at ICM was the catalyst that got the book idea started in the first place and helped appreciably throughout. In the first few weeks, author Bill Barich also helped significantly to focus and articulate our random thoughts into an actual project.

During the long months of writing and research, there were three people who were a steady source of both practical and emotional support for the project. Stanford PhD candidate Siobhan O'Mahoney was both clever and persistent in pursuit of information and supporting evidence. Scott Underwood, who loves words more than anyone I know, helped clarify facts and the nuance of language to describe them. And Joani Ichiki helped make order out of chaos, working patiently through our jumbled combination of e-mails, handwritten edits, and scribbled Post-it notes.

Just when the text was nearing completion came the surprisingly complex task of shooting, gathering, and organizing the images that appear in the final book. Lynn Winter was nearly superhuman in her energy and persistence on that part of the project, allowing me to focus on the written word. Photographers Joe Watson and Steve Moeder shot lots of original photography and IDEO's graphic designer Stephanie Lee helped create some of the composite images. (My kids

insist that I reassure you, no mice were harmed in the elephant photo shoot, as far as we know.)

There were others who helped significantly throughout the process. Whitney Mortimer played a nearly continuous role as a source of business judgment, resource access, and practical advice. Roger Scholl at Doubleday was kind enough to leave us alone during the development of the first manuscript and firm enough to keep us on track as the publication date grew near. IDEO CFO Dave Strong, who sits across the aisle from me, generously looked the other way on days and weeks when writing and revisions seriously impacted my day job as the firm's general manager.

I want to especially thank all the people at IDEO, who have shared their time and their ideas. They were always willing to tell me their favorite stories, answer my e-mail queries, and even hold brainstormers on book-related topics. The list of IDEO contributors here is too long to mention, but you know who you are.

Jonathan Littman, my coauthor, not only shouldered most of the heavy lifting during creation of the first manuscript, but also taught me a lot about writing in the process. I gained new respect for his profession and am anxious to see his future works.

As for my brother David, the dedication does not begin to tell the story. He was—and is—a major influence in my life, and I have never taken for granted the lucky accident of birth that made me his brother. Most of the principles underlying this book came directly or indirectly from David and the work practices he created at IDEO.

To all the rest of my family, thanks for the support, stamina, and love during this long and intense project. My wife Yumi did more than her share of the parenting in the last year, and my two enthusiastic-but-patient kids got good at starting sentences with the phrase "After the book is done, do you think we could. . . .?" As you read this text, I am off with them somewhere, making up for lost time.

CONTENTS

FOREWORD BY TOM PETERS xi

1 INNOVATION AT THE TOP 1

2 WINGING IT IN START-UP MODE 15

3 INNOVATION BEGINS WITH AN EYE 23

4 THE PERFECT BRAINSTORM 53

5 A COOL COMPANY NEEDS HOT GROUPS 67

6 PROTOTYPING IS THE SHORTHAND OF INNOVATION 101

7 BUILD YOUR GREENHOUSE 119

8 EXPECT THE UNEXPECTED 147

9 BARRIER JUMPING 163

10 CREATING EXPERIENCES FOR FUN AND PROFIT 193

11 ZERO TO SIXTY 219

12 COLORING OUTSIDE THE LINES 229

13 IN SEARCH OF THE "WET NAP" INTERFACE 253

14 LIVE THE FUTURE 275

15 GETTING IN THE SWING 291

INDEX 299

It was 1990. I had been a consultant and management lecturer for over fifteen years. Probably hung out in three or four hundred companies. But . . . this one was different.

Just six blocks from my Palo Alto office, I'd never visited it. And now, following a half-day tour, I recall clearly bouncing in the front door of our office and saying to our receptionist, the first person I encountered, "It's finally happened. I've seen a company where I can imagine working!" (In retrospect, I guess that was a frightening thing to say to her.)

The company in question was IDEO (actually, at the time, David Kelley Design). And I'd been bowled over by the spirit and sense of playfulness that invaded every aspect of its stellar—wildly creative—work.

I hope I'm not generally a braggart, but in this instance I claim some precedence. I think I was the first of the "gurus" to latch on to IDEO as Exhibit A in the folder marked "innovation machines."

That was then, and in the subsequent ten-plus years, innovation has spurted to the tippy top of the "requisite core competence list" for companies of all shapes and sizes. And still, nobody does it better than IDEO.

But how? Fat chance of finding out, as IDEO's finely-tuned methodology is obviously its best kept secret.

Until now.

Enter . . . *THE ART OF INNOVATION*

Tom Kelley, IDEO exec and David Kelley's brother, tells all!

This is a marvelous book. It carefully walks us through each stage of the IDEO innovation process—from creating hot teams (IDEO is perpetually on "broil") to learning to see through the customer's eyes (forget focus groups!) and brainstorming (trust me, nobody but nobody

does it better than IDEO) to rapid prototyping (and nobody, but nobody, does it better . . .).

But this is no drab and dreary academic tour. Hey, IDEO creates very cool "stuff" of all sorts. And the case studies—from grocery carts to toothpaste tubes, electronic doodads to obscure medical devices—breathe life into practically every page of the book.

In recent years, as the L.O.I. (Legend of IDEO) has spread far and wide, the company has had clients begging for advice not just on a product or two, but on the IDEO way of innovating. It has responded vigorously. That's good news for readers. It means this methodology not only works for IDEO, but has proven to be transferable.

It's not quite that simple, of course. Beneath the IDEO method lies the incredible, throbbing IDEO spirit that led me to love at first sight. No, it won't be "1, 2, 3 . . . I'm an innovator now." Nonetheless, I can imagine no better launching point than the pages, ideas, and cases of this book. I have been waiting ten years for it. And now I'm lucky enough to own a thoroughly highlighted copy of the galleys . . . that I will barely let out of my sight.

Innovation is it, for the foreseeable future. And *The Art of Innovation* is it for those with the nerve to take the plunge.

So . . . on with the show!

Tom Peters
Buenos Aires
October 9, 2000

Steven Moeder

Innovation wasn't always a hot topic in the Silicon Valley. More than a decade ago, when our firm was just a small group of product designers working over a dress shop in Palo Alto, we became very interested in why companies looked outside for product development. We hired a professional services firm to help answer that question, and after interviewing many clients (and nonclients) we distilled the answers down into four key reasons: One was just raw *capacity*. Companies had a bigger appetite than their in-house resources could satisfy. The second was *speed*. If they couldn't find anybody in-house to sign up to some incredibly tight deadline, they would look outside. The third reason was the need for some specific *expertise* outside their core competencies. And the fourth was *innovation*.

Well, a funny thing has happened in the ensuing years. Innovation has risen from the bottom to the top of the list. During that time, IDEO has broadened its client base to include some of the best-known and best-managed companies in the world. I personally have met with executives from more than a thousand companies to talk about their organizations' emerging technologies, market perceptions, and, of course, product development plans. With more than a thousand firsthand experiences, it's hard *not* to spot emerging trends unless you are truly asleep at the wheel. The biggest single trend we've observed is the growing acknowledgment of innovation as a centerpiece of corporate strategies and initiatives. What's more, we've noticed that the more senior the executives, the more likely they are to frame their companies' needs in the context of innovation.

To those few companies sitting on the innovation fence, business writer Gary Hamel has a dire prediction: "Out there in some garage is an entrepreneur who's forging a bullet with your company's name on it. You've got one option now—to shoot first. You've got to out-innovate the innovators."

Today companies seem to have an almost insatiable thirst for knowl-

edge, expertise, methodologies, and work practices around innovation. The purpose of this book is to help satisfy some of that thirst, drawing on IDEO's experience from the "front lines" of more than three thousand new product development programs. Our experience is direct and immediate, earned from practical application, not management theory. We've helped old-line Fortune 500 companies reinvent their organizations and bold young start-ups create new industries. We've helped design some of the world's most successful products, everything from the original Apple mouse, once called "the most lovable icon of the computer age," to the elegant Palm V handheld organizer. Whether you are a senior executive, a product manager, an R&D team leader, or a business unit manager, we believe this book can help you innovate.

One of the advantages of our front-lines experience is that we've collected a wealth of contemporary success stories from leading companies around the world. We've linked those organizational achievements to specific methodologies and tools you can use to build innovation into your own organization. I think you'll find that this book will help you to arrive at insights that are directly relevant to you and your company.

I joined IDEO in the late 1980s, when it was reaching that critical stage at which many start-ups either stall or implode. Since that time, however, IDEO has grown dramatically in size and influence, and *Fast Company* magazine now calls it "the world's most celebrated design firm." The *Wall Street Journal* dubbed our offices "Imagination's Playground," and *Fortune* titled its visit to IDEO "A Day at Innovation U." Every spring, *BusinessWeek* publishes a feature story on the power of design in business and includes a cumulative tally of firms who have won the most Industrial Design Excellence Awards. IDEO has topped that list for ten years running.

What's unique about IDEO is that we straddle both sides of the innovation business, as both practitioners and advisers. Every day we work with the world's premier companies to bring innovative products and services to market. Even the best management consulting firms

don't enjoy that hands-on, in-the-trenches experience. Yet, like the best consulting firms, we sometimes host teams from multinational companies who want to learn from our culture and steep themselves in our methodology. In other words, we don't just *teach* the process of innovation. We actually *do* it, day in and day out.

As I was completing this book, Tiger Woods was winning the U.S. Open golf tournament at Pebble Beach, dominating the field as never before. He seemed both intense and utterly calm. His dedication was complete, and his swing and putting were nearly perfect. In spite of what looked like masterful putting in his first round, he insisted that the balls weren't going into the hole smoothly enough for him. They were just "scooting," he said, not rolling. He stayed on the practice green till they rolled beautifully. Butch Harmon, his swing guru, said Tiger was playing better than ever. "He's confident. He's mature," said Harmon. "We've built his swing together, so it's pretty easy to tweak if something goes wrong." I found that a wonderful, enlightening statement. The greatest golfer in history, who appears to be the ultimate solo performer, is actually the product of a *team* effort, and when the occasional bumps in the road arrive, the going is easier because of that fact.

Our approach to innovation is part golf swing, part secret recipe. There are specific elements we believe will help you and your company to be more innovative. But it's not a matter of simply following directions. Our "secret formula" is actually not very formulaic. It's a blend of methodologies, work practices, culture, and infrastructure. Methodology alone is not enough. For example, as you'll see in chapter 6, prototyping is both a step in the innovation process and a philosophy about moving continuously forward, even when some variables are still undefined. And brainstorming (covered in chapter 4) is not just a valuable creative tool at the fuzzy front end of projects. It's also a pervasive cultural influence for making sure that individuals don't waste too much energy spinning their wheels on a tough problem when the collective wisdom of the team can get them "unstuck" in less than an hour. Success depends on both *what* you do and *how* you do it.

THE INNOVATION DECATHLON

Here's the good news. Neither you nor your company needs to be best of class in every category. Like an Olympic decathlon, the object is to achieve true excellence in a few areas, and strength in many. If you're the best in the world at uncovering your customers' latent, unspoken needs, the strength of your insights might help you succeed in spite of shortcomings elsewhere. Similarly, if you can paint a compelling visualization of the future, maybe your partners (suppliers, distributors, consultants, etc.) or even your customers can help you get there. If there are ten events in creating and sustaining an innovative culture, what counts is your total score, your ability to regularly best the competition in the full range of daily tests that every company faces.

A METHOD TO OUR MADNESS

Because of the eclectic appearance of our office space and the frenetic, sometimes boisterous work and play in process, some people come away from their first visit to our offices with the impression that IDEO is totally chaotic. In fact, we have a well-developed and continuously refined methodology; it's just that we interpret that methodology very differently according to the nature of the task at hand. Loosely described, that methodology has five basic steps:

1. **Understand** the market, the client, the technology, and the perceived constraints on the problem. Later in a project, we often challenge those constraints, but it's important to understand current perceptions.
2. **Observe** real people in real-life situations to find out what makes them tick: what confuses them, what they like, what they hate, where they have latent needs not ad-

dressed by current products and services. (More about this step in chapter 3.)

3. **Visualize** new-to-the-world concepts and the customers who will use them. Some people think of this step as predicting the future, and it is probably the most brainstorming-intensive phase of the process. Quite often, the visualization takes the form of a computer-based rendering or simulation, though IDEO also builds thousands of physical models and prototypes every year. For new product categories we sometimes visualize the customer experience by using composite characters and storyboard-illustrated scenarios. In some cases, we even make a video that portrays life with the future product before it really exists.

4. **Evaluate and refine** the prototypes in a series of quick iterations. We try not to get too attached to the first few prototypes, because we know they'll change. No idea is so good that it can't be improved upon, and we plan on a *series* of improvements. We get input from our internal team, from the client team, from knowledgeable people not directly involved with the project, and from people who make up the target market. We watch for what works and what doesn't, what confuses people, what they seem to like, and we incrementally improve the product in the next round.

5. **Implement** the new concept for commercialization. This phase is often the longest and most technically challenging in the development process, but I believe that IDEO's ability to successfully implement lends credibility to all the creative work that goes before.

We've demonstrated that this deceptively simple methodology works for everything from creating simple children's toys to launching e-commerce businesses. It's a process that has helped create products that have already saved scores of lives, from portable defibrillators and

better insulin-delivery systems to devices that help grow sheets of new skin for burn victims.

INNOVATING WITH AN AUDIENCE (AND WITHOUT A NET)

Part of the reason this book came about is that we actually got first-hand evidence that people believe in our approach to innovation. A year ago, ABC's *Nightline* came to us with a unique proposition. They wanted to "see innovation happen" and said that, if we were willing to show how we'd reinvent a product category, *Nightline* would be there with its cameras to capture the action.

Perhaps you're one of the almost 10 million people who stayed up late to watch the broadcast.

The show was great entertainment, but it was also a wonderful short course in our methodology. ABC had asked us to compress our method for creating successful products into a TV-sized package, and the steps we went through before a national television audience are the very steps that I'll take you through in the rest of this book. As a preview, let's dive into what ABC called "The Deep Dive. One company's secret weapon for innovation."

THE DEEP DIVE

Nightline's show began with Ted Koppel asking how the process of designing a better product works. He went on to describe the toughest problem the network could think of to toss our way. "Take something old and familiar," he said. "Like, say, the shopping cart, and completely redesign it in just five days."

That's exactly what we did.

"Maybe we should acknowledge it's kind of insane to do an entire project in a week," began Peter Skillman of IDEO as ABC's cameras

rolled. It was 9:00 Monday morning, day one, and the youthful Skill-man was a walking metaphor for the IDEO way. Status at IDEO is about talent, not seniority, and Skillman had proved an able facilitator under fire, great at leading brainstorms and bringing disparate teams together. The team that day at our Palo Alto offices also came from many disciplines. Beyond our usual talented engineers and industrial designers we had IDEOers with backgrounds in psychology, architecture, business administration, linguistics, and biology.

The shopping cart was an ideal and imposing challenge. The cart is an American cultural icon, as familiar as the Zippo lighter, and just as equally frozen in time. It offered a rich opportunity for new design, but at the same time we knew that it was inexplicably stuck in a sort of in-novation limbo.

"Let's go!" Skillman cheered at 10:00 A,M., and we were off and run-ning. We split into groups to immerse ourselves in the state of grocery shopping, shopping carts, and any and all possibly relevant technolo-gies. Blending our "understand" and "observe" phases into a single day's work, we were practicing a form of instant anthropology. We were getting out of the office, cornering the experts, and observing the na-tives in their habitat. Some members of the team trotted down to Whole Foods, a popular grocery store in downtown Palo Alto, and be-gan wandering the aisles, watching with a fresh perspective how people shop. They saw safety issues and watched parents struggle with small children. They noted how professional shoppers from an Inter-net buying service used their carts as a base station and ran up and down the aisle "cartless" for better mobility. They saw traffic jams where shoppers had to pick up the back of their carts to slide by other slow-moving or oncoming ones.

I interviewed a professional buyer who purchases carts for a large store chain, and discovered the trade-offs of steel versus plastic, as well as the surprisingly high cost of lost and damaged units. Another group caught up on the latest designs and materials by cruising a local bike store. A "family" team poked and prodded a dozen children's car seats and baby buggies. Anticipating that we'd "cyberize" our cart, we

perused a local electronics store for gadgets. One group managed to track down a cart repairman named Buzz who drives from Safeway to Safeway in a little truck, rewelding broken baskets and popping on new wheels.

By the end of day one, three goals had emerged: make the cart more child-friendly, figure out a more efficient shopping system, and increase safety.

Focusing on those themes, we spent the morning of day two brainstorming possible solutions. The classic brainstorming principles were printed on the walls, and we spread giant Post-it sheets with lots of colored markers about and plenty of toys to lighten the mood. We didn't fret if an idea was dull or even goofy, and we encouraged everyone to join the show-and-tell. The wacky concepts cracked everybody up and kept people from editing their own thoughts, like the privacy shade someone sketched (in case you're buying six cases of condoms) or the Velcro seats with matching Velcro kid diapers to keep unruly toddlers safely stuck in place.

By 11:00 A.M. the focused chaos started winding down, hundreds of crazy ideas and sketches crowding the walls, as well as plenty of solid ones, like a cart that nobody would want to steal or a cart with its own scanner to check prices. We voted for the "cool" ideas. They couldn't be too far-out, because they had to be buildable in a couple of days. Everyone stuck brightly colored Post-its on their favorites, creating flowerlike clusters around the best concepts.

Over lunch, the team leaders reviewed the concepts and the group's votes and made a series of quick decisions on where to focus prototyping efforts. The fastest development teams in the world can't win the race to market if the decision process bogs down, so by the time the pizza was finished, the Deep Dive team had a plan for going forward.

We split into four smaller groups that would have three hours to build mock-ups, each team focusing on a separate concern—shopping, safety, checkout, and finding what you're looking for.

The groups sketched their ideas for half an hour and then took off

running. Many jammed the aisles at the local Ace Hardware store searching for ideas and materials. One of our master model makers pursued an idea from the brainstorm to make a shopping cart that tracks sideways. By 3:00 P.M. on Tuesday, sixteen IDEOers were jammed into our shop along with the dozen machinists and model makers who work there every day. They were feeling the intense time pressure to crank out the first round of sample shopping carts, and three hours later the crude prototypes were ready for review. One featured an elegant and voluptuous curve; another was modular, designed to stack up with handbaskets. There were high-tech twists—a microphone to query customer service and a scanner so you could skip the checkout line.

Again, we selected the best feature of each prototype and divided the tasks. Next, it was Lego time—everyone started bending wire-welding rods to build tiny model carts. We knew it had to be modular, child-safe, and nestable for easy storing. While one person was laying out the frame assembly on a CAD machine, someone else was examining the basket concepts. The design team called it a day at 3:00 in the morning, but the shop kept at it a little longer.

At 6:00 A.M. Wednesday, day three, a master welder whom IDEO works with picked up the drawings for the tricky, curvaceous frame. Meanwhile, model maker Jim Feuhrer was tinkering with the casters. The incredibly challenging deadline and shared goal had helped create a spirited "hot group," and the team pushed through another long day, fueled by energy reserves and nearby Peet's espresso. By Thursday afternoon the team was getting punchy, but beginning to think it was possible. My brother David, founder of IDEO, came bounding through the shop with his usual infectious optimism and told everyone, "It looks great. It's awesome." They'd started to put the parts together and had Whole Foods baskets all set up to insert into our custom frame. David's expression suddenly changed. "You're not going to use *those?*"

The team had focused so completely on redesigning the shopping

cart *frame* that they hadn't had time to work on the *baskets*. Only hours remained, but shop leader Carl Anderson and others grabbed some acrylic panels and started cranking. Meanwhile, at every point, test assemblies were being done. The shop kept at it nearly till dawn. But the cart still wasn't done. We had to paint it before the cameras were ready to roll a few hours later.

LIFTING THE CREATIVITY CURTAIN

At 9:00 A.M. on Friday we wheeled the cart down the street, put it in a conference room, and threw a sheet over it. Everyone gathered round for a cheer as we yanked off the sheet to a television audience of millions.

ABC loved what it saw. The old boxy cart we all know and hate had been replaced by a sleek, gleaming creation. The main frame sloped down on each side into a curve that tucked back, with more of a sports car line. Gone was the main basket—the feature that made carts desirable for black-market barbecues. The open frame was designed so that six standard handbaskets would neatly nest inside in two layers. Shoppers could use the cart like home base, darting down an aisle with a basket. At checkout, clerks would pack the groceries in plastic bags that neatly hook within the frame. As far as we know, no one had done anything quite like it before. To me that's the heart of it, a real innovation that redesigns the shopping experience.

We used ideas from roller coasters and baby seats to create the cart's child seat—it had a safety bar that snaps in like one at the amusement park as well as a fun blue plastic play surface. There was a scanner to pay for items directly, two cup holders for coffee, and a clever set of back wheels. Tug to the side and the locked wheels would pop loose so you could easily push the cart sideways. Push it forward again and the wheels would lock back in place.

With the cameras rolling, ABC's Jack Smith wheeled the cart down the aisles at Whole Foods and earned plenty of gawking looks. Clerks

and managers loved the cart and even said that with a couple of modifications, they'd want one. We took the afternoon off to celebrate and get ready to return to our regular clients.

The cart was done, the show was aired, and we thought that was pretty much the end of it. But the morning after the *Nightline* segment ran, our phones wouldn't stop ringing. I took dozens of calls from executives around the country who'd seen the show. Most of them didn't give a damn about shopping carts. Instead, they wanted to know more about the process we used to bring the cart into being. One CEO told me that he understood, for the first time, what creativity really meant and how it could be managed in a business environment.

Nightline's Deep Dive broadcast was among its most popular of the year, so popular in fact that the network rebroadcast it a few months later. The response amazed us. But maybe it shouldn't have. The fact is, everybody talks about creativity and innovation, but not many people perform the feats without a safety net in front of a nationwide television audience.

BUILDING IN CREATIVITY AND INNOVATION

Why should business care about creativity? Visit your local mall or trade show and you'll see that creativity sells. We're all searching for the next iMac or VW Beetle—any worthwhile innovation that captures the public's imagination and strengthens the company's brand. But many companies shy away from novel solutions. Moreover, they tend to believe that truly creative individuals are few and far between. We believe the opposite. We *all* have a creative side, and it can flourish if you spawn a culture to encourage it, one that embraces risks and wild ideas and tolerates the occasional failure. We've seen it happen.

The more we thought about the success of the *Nightline* Deep Dive,

the more it made sense to distill what we've learned in the trenches from hundreds of corporations on thousands of projects. This book aims to demystify the creative process. It isn't something we dreamed up in a business school class. It's been tried and tested through hands-on experience. It helped IDEO grow from a two-person office into the leading product design firm in the world.

It can help you too.

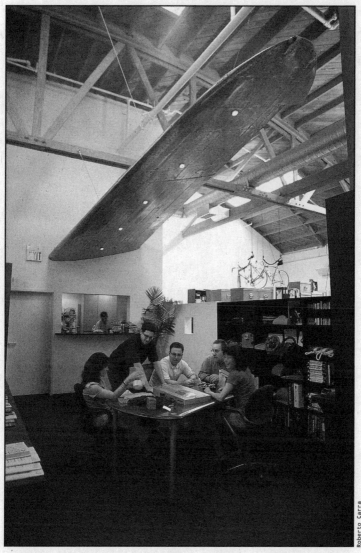

When my brother David launched his business in 1978, he treated it as a project, like the snow forts of our childhood or the thirty-foot jukeboxes he made in college. The truth is, he just winged it and had a lot of fun in the process.

After less-than-exciting engineering roles at Boeing and NCR, David found his spiritual home in the Product Design program at Stanford, where he earned his master's degree in the late seventies. Reluctant to leave such a stimulating and nurturing environment, David thought he'd go for a doctoral program. But the Ph.D. involved heavy doses of reading and writing—not my restless brother's favorite pastimes. When the Stanford Design Department kept getting inquiries from companies looking for someone to solve their tricky engineering and product design problems, David decided he could do it.

David recognized that he had to start his own business. He knew he couldn't fit in a conventional workplace. He wasn't linear. His forte wasn't sitting down and working. Nor did David have the attention span to follow somebody else's direction. If he wanted to have a chance at success, he'd have to lead.

David decided he needed a partner and asked his Stanford mentor, Professor Bob McKim, for the name of a top student in that year's program. McKim suggested Dean Hovey, and after Dean agreed to partner with him, David promptly abandoned his plans for a Ph.D. The first four engineers they hired—Jim Yurchenco, Dennis Boyle, Rickson Sun, and Douglas Dayton—were all Stanford grads who were friends of David's. David believed that if he hired people he liked and respected, everybody would have fun and get more work done.

The work was truly like child's play—they made things up as they went along. They found a couple of rooms at a dilapidated Palo Alto office above a dress shop. They made most of their own furniture. They spray-painted used chairs bright green and laid discount doors from the local lumberyard over dented filing cabinets to make desks. To

match the chairs, they bought some green, cheap indoor-outdoor carpet. Finally, they slapped some paint on the walls and nailed up some sheetrock dividers.

The room dividers were intended to give everyone some semblance of privacy. But Dennis Boyle promptly cut a circular hole in the wall between him and Douglas Dayton and popped in a ship's porthole. Adversity brought them together. The musty offices were so thick with flies that Yurchenco and Dayton built a funnel out of foam core to suck up the insects with a vacuum. They called themselves the Fly Group. For kicks, they built a fly the size of a pig out of foam core, painted it red, and hung it from the ceiling. Pranks became second nature. When Hovey left for a week's vacation, he returned to find a sheetrock wall where his door had been.

Windshield cement inspired many office pranks. You'd leave your desk only to return to find everything glued down: soda cans, papers, pens. David's door was once glued shut when he was getting a pitch from a salesperson. Another office was webbed in by the sticky trails from a hot glue gun. There were rubber band wars and squirt gun skirmishes (similar to the pranks at Apple at the time), and plenty of water balloons dropped out the window. Someone even built a key-cap launcher with a bucket of keyboard keys left over from a job. IDEO was like hanging out with your friends on summer break.

Jim Yurchenco says the pranks and play served a purpose—gave him a sense that he had some control over his destiny, a feeling of belonging to something larger than himself. Something of a loner, he was anything but a typical hire. Yurchenco had never studied product design—or engineering, for that matter. He had earned a master's in fine arts from Stanford and spent most of his time in the university's design shop making arcane electromechanical sculptures. It sounds loony, hiring a sculptor for a product design firm, but David was no fool. Yurchenco had excelled in math and physics and, like David, had grown up building things. David knew that Yurchenco was wildly capable.

The company had no business plan. David hustled up jobs. One of

the company's neighbors in downtown Palo Alto, Jerry Mannock, was an industrial designer and fellow Stanford grad. David dropped by to tell him about the new firm. Mannock happened to be doing a lot of work for Apple at the time, and next thing you know, David was going down to meet Steve Jobs.

David didn't have a lot of experience, but neither did anyone else, it seemed. Companies big and small in the Silicon Valley were turning to people like David as they struggled to solve novel engineering and design problems in computers and other high-tech devices. Jobs asked David and his newly-formed team to help create the Lisa computer—forerunner to the Mac—as well as the mouse that would control its innovative graphic interface.

Like the e-commerce revolution twenty years later, it was a time when being old and wise wasn't much of an advantage. You had to track down sources that could help you, and be bold enough to make some educated guesses. As David says, "When you're stuck with a tough decision or a problem you don't understand, talk to all the smart people you know." It's the networking approach to problem solving, a lesson he learned in the early years of the firm.

It didn't hurt that David was hanging out with some groundbreaking companies. Apple's confidence was infectious. David loved Apple's hipness, the fact that you could wear jeans and pad about in your stocking feet. He left Apple meetings feeling jazzed by its culture of innovation, by the way Apple's labs and offices intertwined. Workers of all ages and experience seemed to effortlessly cross-pollinate. A teenager might be working next to a veteran engineer from Hewlett-Packard. There was a swagger in the air, the sense that Apple could take on any challenge and succeed. David heard tales of Jobs giving a block of stock to a draftsman, taking everyone to *Star Wars* in the middle of the day, quitting early for a volleyball game. And yet, *they got things done.*

David didn't exactly follow the rules himself. There wasn't room for a machine shop, so he got some two-by-fours and lots of corrugated fiberglass and covered up a central atrium. The floor was the asphalt

The Apple mouse was a breakthrough innovation that became an enduring icon of the computer age.

roof of the dress shop below, and there was no door. To get into the shop, you had to climb out a window, pulling a power cord with you for juice. This makeshift prototyping space probably violated a dozen building codes and two dozen fire codes. But it was a quick and dirty place to stick a couple of saws and a drill press and other tools.

The boyish pranks and wild play didn't just pump up the team. They also created an atmosphere where you naturally took chances and solved problems. You could stumble, as long as you fell forward. The team chalked up its share of blunders: parts that didn't fit together, computers that wouldn't pass muster at the FCC, mirror-image part

drawings. But the team picked itself up, absorbed the hard lessons, and went on.

David was amazed at the opportunities. He'd expected to be kept behind the scenes—engineers at places like Boeing and NCR were treated merely as cogs in the wheels of industry—but within a couple of years, David was working for major corporations and meeting with company presidents. In high tech the executives really cared about products and about innovation.

Slowly, David's company emerged from its first big transition. David's partner traded in his share of consultancy for the majority ownership of a spin-off manufacturing arm. But the workers opted to stick with David and what became David Kelley Design.

Ten years later, responding to client requests for "one-stop shopping," the company went through another major transition by combining forces with Moggridge Associates in London, ID Two in San Francisco, and Matrix in Palo Alto. In casting about for a name for the new company, Bill Moggridge picked the prefix *ideo* out of the dictionary (as in "ideology"), and IDEO Product Development was born in 1991. Since that time, the firm has steadily grown and diversified, without ever seeing an unprofitable quarter.

Along the way, we've played a role in many key developments, everything from mobile computing and Internet appliances to minimally invasive surgery and cardiovascular monitoring.

As we've built expertise and credibility in some areas, we continue to "wing it" with new experiments in alternative business models, international locations, and innovative service offerings. IDEO continues to be a work in progress.

We wouldn't have it any other way.

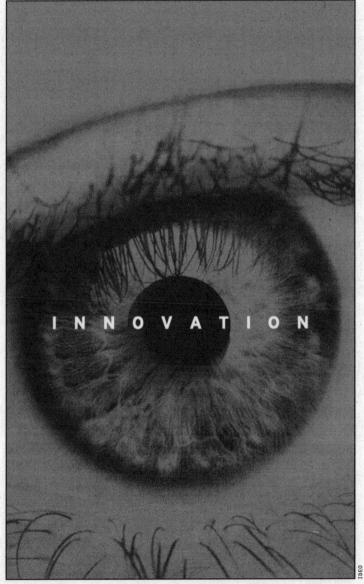

Whhat do stand-up toothpaste tubes, all-in-one fishing kits, high-tech blood analyzers, flexible office shelves, and self-sealing sports bottles have in common? Nothing actually, except that they're all IDEO-designed products that were inspired by watching real people.

We're not big fans of focus groups. We don't much care for traditional market research either. We go to the source. Not the "experts" inside a company, but the actual people who use the product or something similar to what we're hoping to create.

Plenty of well-meaning clients duly inform us what a new product needs to do. They already "know" how people use their products. They're so familiar with their customers and existing product line that they can rattle off half a dozen good reasons why an innovation is impractical. Of course, we listen to these concerns. Then we get in the operating room, so to speak, and see for ourselves.

A few years back, for example, Silicon Valley–based Advanced Cardiovascular Systems asked us to help it redesign a critical medical instrument used on heart patients during balloon angioplasty. The company sold an inflation device for the tiny balloon that the doctor inserts with a catheter through the femoral artery in a patient's leg. The balloon is guided up into the obstructed coronary artery and inflated, compressing the plaque and stretching the artery. ACS told us that the new inflation device—like the existing one—had to be suitable for one-handed use.

But when we went into the operating room—literally—that's not what we saw. Although the current product could theoretically be used with one hand, it really worked that way only if you had a hand the size of Michael Jordan's. In actual practice, medical technicians almost always used *both* hands with the device, since, as we observed, they weren't doing anything else with their "spare" hand. So why not design

the new "Indeflator," we thought, for a two-handed technician? Why fight human instinct?

It's precisely this sort of observation-fueled insight that makes innovation possible. Uncovering what comes naturally to people. And having the strength to change the rules. From the simple observation that technicians used *both* hands flowed distinct improvements. We added ribs to the base of the pumplike device so that technicians could hold it steady in one hand while they inflated the balloon with the other hand. We tilted the pressure gauge upward so that it was easy to read during inflation. We increased control and precision. We made it easier to deflate the balloon too. And we made one other big change.

There's a critical moment in an angioplasty procedure when the surgeon instructs a technician to inflate the balloon. During the next sixty seconds or so, the balloon obstructs the artery, creating, in effect, a heart attack. At that point, with the patient still awake, the old device would make a loud clicking noise as it ratcheted into place.

Our new design lost that scary ratcheting sound.

TIME IN THE JUNGLE

Clicks-and-mortar brokerage founder Charles Schwab has talked about his effort to assume the perspective of his customers. "I am like a chef. I like to taste the food. If it tastes bad, I don't serve it. I'm constantly monitoring what we do, and I'm always looking for better ways we can provide financial services, ways that would make me happy if I were a client."

Noble aspirations, and you can't argue with Schwab's track record, but we believe you have to go beyond putting yourself in your customers' shoes. Indeed, we believe it's not even enough to *ask* people what they think about a product or idea.

One reason is the same factor that prevents you from learning that

your meat loaf tastes like sawdust. Your dinner guests are too polite to tell you the unvarnished truth, too wrapped up in trying to give you the expected answer. How's the meat loaf? "Fine," they say. ("Delicious," if they care about you or think it will make you happy.) How many people volunteer that they're having a lousy day? It's human nature to put a bright face on a dismal situation. Because there's no information, no value, no content to the "fine" response, we sometimes say, "Fine is a four-letter word."

A second reason for the "fine" response is that your guests don't know or can't articulate the "true" answer. Maybe the meat loaf needs more salt or less onion. The problem is that your guests may like to eat, but they're probably not food critics. In business, too, your customers may lack the vocabulary or the palate to explain what's wrong, and especially what's *missing*.

Companies shouldn't ask them to.

This is particularly true of new-to-the-world products or services. A user of a new type of remote control may not be able to recognize that it has too many buttons. Inexperienced computer users may not be able to explain that your Website lacks navigational clues. And they shouldn't have to. We saw this firsthand when a software company asked us to find out how users would react to one of their new applications. We set up a few computers and observed people struggling with the program. More than a couple were having a terrible time, grimacing and sighing audibly as they fumbled with the keyboard and mouse. But in exit interviews, the software company was given a different story. Those same people swore that they'd had no trouble with the new application and couldn't imagine a single improvement.

Customers mean well—and they're trying to be helpful—but it's not their job to be visionaries. Indeed, former 3Com CEO Bob Metcalfe tells the story of how, in the early eighties, his customers and salespeople practically *demanded* that he dedicate their R&D efforts to making a new version of its networking card for multibus-compatible computers. Metcalfe balked, and some of his salespeople quit in protest, disgusted that the company seemed to be ignoring the re-

quests of its own customers. Instead, 3Com chose to develop an EtherLink card that worked with the new IBM PC. Today there are *no* multibus computers left in the world, but 3Com ships more than 20 million EtherLink cards a year.

Seeing and hearing things with your own eyes and ears is a critical first step in improving or creating a breakthrough product. We typically call this process "human factors." I prefer "human inspiration" or, as IDEO human factors expert Leon Segal says, "Innovation begins with an eye." It's a general principle of humankind. Scientists, industrialists, anthropologists, artists, and writers have understood this for centuries, and many entrepreneurs understand it intuitively.

Once you start observing carefully, all kinds of insights and opportunities can open up. For example, the hugely popular elliptical cross-trainer exercise machines in your local health club got started from a simple human observation. Larry Miller, a human-factors-savvy person working at General Motors, was videotaping his daughter running one day and noticed the elliptical path traced by her feet as she went through her exercise. From that observation-based spark, Miller set about building a prototype of a device that would mimic his daughter's elliptical movement—without the jarring impact of feet hitting the ground. He sold his idea to Seattle-based fitness equipment maker Precor, Inc., which developed it into its EFX line of elliptical trainers. Thanks in part to Miller's epiphany, Precor is now the fastest-growing equipment company in the health club industry.

NETTING A BUG LIST

Sometimes—if you're lucky—you can find inspiration for innovation by observing yourself. In many parts of your life, you go through steps so mechanically, so unconsciously, that this is not possible. When you're off your own beaten path, however, you are more open to discovery: when you travel, especially overseas; when you rent an unfa-

Daniels & Daniels

Larry Miller got the inspiration for his elliptical trainer
by watching his daughter run.

miliar car; when you try a new sport or experience a new activity. At those times, you are more open to ask the childlike "Why?" and "Why not?" questions that lead to innovation. Whenever you are in that new-to-the-experience mode, I would urge you to pay close attention and even take notes about your impressions, reactions, and questions. Especially the problems, the things that bug you. We call these mental and jotted-down observations "bug lists," and they can change your life. That's what happened one day to twenty-six-year-old Perry Klebahn on a visit to a Lake Tahoe ski resort.

Klebahn was recovering from an ankle fracture, and although he could walk without pain, his doctor had warned that skiing was inadvisable. Still wanting to meet his friends for lunch on the slopes, Klebahn discovered that the resort had some snowshoes available to help him traverse the snowy terrain. Using snowshoes for the first time, he was struck by how incredibly awkward they were to use. For one thing, they weighed more than ten pounds, turning what would have been a pleasant walk into serious exercise. On level or uphill terrain, the front of the snowshoes would fill up with snow, making them even heavier and causing you to trip over your own feet. Whenever there was a downhill slope, the shoes were hard to control and would sometimes slip out from under you. All in all, a pretty unpleasant experience, and a product category that had not seen much innovation since Lewis and Clark. A fatalist would have just written off snowshoes as awkward, antiquated equipment, but Klebahn was a Stanford product design student at the time, learning how to sharpen his observation skills, keeping bug lists, and asking a lot of "why?/why not?" questions.

Inspired by observing his own difficulties with the existing technology, Klebahn—while still a student—formed Atlas Snowshoe Company, which almost single-handedly created today's snowshoe industry. Using a clever design and high-tech materials, he cut the weight of the snowshoes by 70 percent and made them easy to use on any terrain. That left the small task of creating an industry around his new product, but within a few years, Atlas had partnered with ski resorts from Vancouver to Sun Valley in creating snowshoeing areas. Resorts initially worried "if we build one, will they come?" but a single snowshoe area at Vail boasted more than 100,000 visitors by its second season. Perry Klebahn, starting with a single observation, then following up with a lot of creativity and hard work, grew Atlas Snowshoe Company to more than $10 million in sales and then sold the company.

Anecdotal? Yes, but hardly an isolated case. Ask around, and you'll find that many entrepreneurs got started by observing humans struggling with tired routines and asking themselves what they could do about it. Scott Cook, cofounder of Intuit, got the idea for the com-

pany's first product by observing his wife paying the bills in the slightly tedious manual way. He wondered whether there wasn't a way to "quicken" that process. Intuit had sales of almost a billion dollars in 1999, and Quicken is still the company's most successful product.

KEEPING CLOSE TO THE ACTION

Whether it's art, science, technology, or business, inspiration often comes from being close to the action. That's part of why geography, even in the Internet age, counts. And why so many high-tech companies have emerged from Silicon Valley—and not Connecticut or even New York. New ideas come from seeing, smelling, hearing—being there.

This sensory immersion is why people still fly to other parts of the country for face-to-face meetings with clients, customers, and colleagues, even in the information age; why phone or videoconferencing often doesn't do it. It's also why people still go to museums, to be inspired in the presence of original artwork, though a digital image may be easily available on their home computer screen.

Asking questions of people who were there, who should know, often isn't enough. It doesn't matter how smart they are, how well they know the product or the opportunities. It doesn't matter how many astute questions you ask. If you're not in the jungle, you're not going to know the tiger.

NO DUMB QUESTIONS

"She has a good eye for business" is a cliché.

But clichés usually have a grain of truth. The reason people talk about the importance of having "a good eye for business" is that you

need to be aware of the world around you, ready to spot trends—and act. You can't wait for a report or rely on reading it in the paper or on the Web.

Good companies and good consultants are astute observers, of people, teams, organizations, technologies—and trends. They see quirks and patterns. Lots of folks try to do this. What makes IDEO different is that we put a lot of steam and spark behind our observations. And we have come up with some pretty good methods that increase the quality of our observations.

It's a funny paradox. Though we're pretty confident in our ability to observe people and draw insights out of them, we pride ourselves on starting every project humbly—and a little dumb. We don't want to peek at the answers before we know the questions. Steelcase, the world's leading manufacturer of office furnishings and an IDEO investor, asked Sean Corcorran's team at IDEO to dream up some concepts for "active storage" as part of a new modular system of interior architecture called Pathways. Sean was enthusiastic but wasn't sure where the exploration would lead. He asked a few questions, and the discussion moved toward "new and emerging work processes."

Of course, Steelcase would seem to know everything there is to know about office environments. But even after the company gave us some direction and shared some of its knowledge, we didn't feel like we had enough intuitive understanding of the issue to start designing. The problem, as Sean recalls, was "still a little too fuzzy." We encounter that a lot: the fuzzy problem versus the well-defined problem. So Sean started with the basics, sending a team of people into offices to take photographs, watch, and ask questions.

If you ask them, most people will insist that they file in filing cabinets or on their computer (remember the promise of the paperless office?). But observe people at work, and you'll find something less idealized. Sean found that people stored paper on the floor, on chairs, on top of binders, and, of course, on their desks or other work surfaces. "Filing by piling" was the phrase that came to mind. Sean's team was on the brink of a valuable insight. If you clutter your desk, where do you work?

The proverbial light switched on. People store on their desks anyway, so why not just create a little desk above the desk? They can still file by piling, but the desk and work space would be less cluttered. And consistent with the Pathways emphasis on mobility, the desk surface itself could be wheeled wherever a worker needed it.

The resulting "Datum shelf" helped shrink the footprint of an office. Steelcase patented the idea and built on it, introducing complementary shelves for office corners or other unused spaces, bringing a deeper three-dimensional geometry to the workplace. Best of all, the Datum shelf celebrated human behavior. Instead of trying to change the primal need to file by piling, we channeled that urge into a more productive pattern.

A CHILD'S EYE

We're big advocates of a principle we call "being left-handed," developing empathy for consumers' needs, even if those consumers are very different from yourself. By studying people of all ages, shapes, cultures, and sizes we've learned that the best products embrace people's differences.

Take kids, for instance. Today the best companies recognize the value of talking—and listening—to kids. More and more companies are putting their products into the hands of teenagers, asking them to give them a test drive and report back. But we look at this secondhand data collection as better for marketing than for inspiration. We prefer getting kids down on the test track and watching them take prototypes out for a spin.

An interactive media group at Philips asked us several years ago to come up with a new computer input device for three-to-five-year-olds. Having had home-run successes with computer mice for Apple, Microsoft, and Logitech, we worked up some mouse prototypes and let

the kids at them. Trouble is, they drove the mice like cars. It was hard enough for them to understand that moving the mouse forward translated to moving the cursor upward on the screen. Worse yet, the kids wanted to turn the mouse ninety degrees and "drive" it left, which, unfortunately, does not move the cursor left at all. Finally, no young child that I have seen using a mouse for the first time can grasp the idea that if you run out of "mousing" space at the edge of the table, you just pick up the mouse and move it a little the other direction.

Insights like these led us to leave the mouse behind and prototype a softball-sized trackball. Big enough to accommodate kids' less developed dexterity, better suited to survive the abuse kids dish out, and easier for youngsters to make that big conceptual leap—if you move the ball, that little whatchamacallit on the screen moves too. All too often, companies overlook this critical step of watching real people. To make better products and services, you've got to care about the person actually using it.

Take something as basic as a toothbrush. The most important time to learn to brush thoroughly and regularly is as a child, yet for decades kids' toothbrushes were pretty much just smaller versions of adult brushes. On a project for Oral-B, we put brushes in the hands of young kids and quickly noticed the "fist phenomenon." Little kids grip the brush with their whole fist, unlike older kids, who use their fingertips. At first, it seems almost paradoxical that kids' toothbrushes would be fatter than adults', but not when you see them in use. So we made a fat, soft, squishy grip that would be easy for them to handle. Oral-B's new Squish Grip brushes looked and felt like toys, a good thing when you consider that the longer kids brush, the better.

This concept of "being left-handed" can apply to all aspects of diversity in your customer base. Kids are obviously different in size, mental development, and attention span. Other groups have subtler differences you need to understand, observe, and develop empathy for if you want to anticipate their interests and needs.

Steven Moeder

Observations helped us discover that smaller
hands actually need fatter toothbrushes.

INSPIRATION BY OBSERVATION

Open your eyes and you'll be awakened to opportunities to improve products and services without even leaving your office. Walking around Bank of America's corporate headquarters in 1989 on a project called Objects for a Better Day at Work, we saw something we hadn't expected. Desks, computers, monitors, and keyboards were state-of-the-art but still didn't allow much leeway for physical differences. If

you weren't the "ideal user," you usually had to jury-rig your setup to fit your body. But in the heart of corporate America, we hadn't expected piles of phone books. We first thought it was another storage problem but soon determined that some workers at the bank were placing phone books under their desks because they couldn't comfortably rest their feet on the floor when seated. To achieve the right height for their keyboard and monitor, their legs were left dangling. It seems like a minor insight. But seeing those phone books firsthand helped inspire a range of products, the most obvious, of course, being a simple adjustable footrest.

That's another example of "being left-handed." Not everybody's like you. Not everybody reads the manual or follows directions. Not everybody is a thirty-year-old six-foot-tall white male. Awaken your antenna to the endless variety of human nature, and you're bound to make customers happier and find new markets.

Baxter Healthcare, for instance, asked us to help create a user-centered design for a very special system that helps patients with heart disease. Known as a left ventricular assist system, it's a life-critical device that keeps your heart pumping while you're waiting for a transplant. Although the external unit weighs only a few pounds, that can still be a burden when you consider that the patient using it may be in a weakened state. So we prototyped and observed, looking for a better way to carry the unit. We tried backpacks and vests. In the end, we discovered that a strap around the waist made it easier for the user to interact with the device. And we designed some simple audible tones that would let people know if the battery was running low.

There's an urgency to medical products. Sometimes the right design can be a matter of life or death.

EMBRACING YOUR CRAZY USER

Focused observation can be a powerful source of innovation. As you observe people in their natural settings, you should not only look for the nuances of human behavior but also strive to infer *motivation* and *emotion*. Good, insightful observation combines careful watching with occasional well-chosen "why?" questions to get at the underlying psychology of a person's interactions with products and services.

When it comes to observation, Jane Fulton Suri of IDEO may be a natural. A British transplant who works out of our San Francisco office, Jane heads many of our human-factors investigations.

Jane is a bird-watcher with an attitude, except the birds she specializes in are humans. As a student of psychology at the University of Manchester, Jane became fascinated with how the physical layout of space around the university influenced the quality of human exchange. Puzzled at how an awkwardly designed building seemed to create intersection points where people ran into one another, Jane launched her first formal human-factors study. She began tracking where and when people congregated, creating maps and snapping photos.

It wasn't long before she turned her college fascination into a budding career consulting in traditional ergonomics for British government agencies and corporations. She studied why rail workers fail to hear the train's siren, why drivers don't spot motorcycles until it's too late, and why circular saws occasionally slice into fingers. The solutions she proposed were critical, but it was always about avoiding catastrophe, about eliminating a negative, never about creating a positive. Requiring motorcycle drivers to turn on their headlights did make a small dent in the number of accidents, but in general, bureaucratic regulations to control the design and usage of things didn't seem to be the answer.

Today at IDEO, Jane works on fixing a lot of product problems *before* they get into consumers' hands. She helps companies plumb people's latent needs and wishes, in order to redesign existing products or find inspiration for entirely new ones. She's part anthropologist and

seer, as well as a good old-fashioned foreign correspondent—just the sort of skills that businesses need to help them discover how their products and services can better be used by their customers.

Jane's work is a mixture of hyperobservation and synthesis. We have no time for detailed scientific studies at IDEO, nor does most of the rest of the business world. We aren't interested in hundreds of carefully qualified users filling out detailed forms or sitting in focus groups. Instead, we usually track down several interesting people to observe and talk to. E-mail is often a good source for finding people. We'll blast out a query to see if anyone knows friends who fit a certain profile or who might let us watch them using an existing product or service.

Jane tries to get under people's skin to figure out *what* they think and do, as well as *why*. It was Jane who analyzed the human factors of a major undertaking by Japanese electronics giant NEC to explore fresh possibilities in its computer line.

NEC did fabulously well in Japan with desktop machines, but for some reason hadn't enjoyed the same success with notebook computers. The company decided to throw its weight behind a new notebook effort and hired us to help. Jane went to Tokyo to shadow NEC computer salespeople. Several facts became immediately apparent. First, Japanese businesspeople loved notebooks partly because space was precious. A family's home computer, for example, was often a notebook machine. And salespeople rarely had a desk, let alone an office. They were always juggling competing needs, and since they usually lacked a desktop computer, the notebook had to do it all. At the office, they expected their notebook computer to include a floppy drive; on the road, battery life was critical. But the objective—making a successful presentation to a client via their laptops—wasn't easy. As a salesperson would pull up a chart, he'd have to awkwardly turn the machine (while still trying to hit the right keys) to show the screen to the customer.

From these critical observations came the inspiration for NEC's design of the Versa product line. The floppy drive could be swapped out for another battery when the salesperson took to the road. The salesperson could pull up a chart and deftly swivel the screen so the client

could see. In the end, the Versa won international awards, was prominently featured in *BusinessWeek*, and, most important, doubled NEC's U.S. market share in its first six months.

The Versa was the first of several successful NEC projects inspired by a collaborative process we called "Greenhouse." One reason it was so successful was that NEC took a holistic multidisciplinary approach to designing its new products.

What does this mean? Many companies rigidly separate functions such as research, design, marketing, and manufacturing, creating walls between groups that have much to teach one another. NEC set out to integrate the whole process, inviting the marketing and manufacturing departments to inform the design and broaden communication. As Jane puts it, "You don't just send your researchers out to do research and your designers to do design. You send your designers with researchers to do design and vice versa."

Jane has spearheaded numerous IDEO efforts to ensure that both designers and clients are part of the observation process—that the discovery process is organic—because it's not enough to see or hear what people say. You have to interpret and intuit shades of meaning to divine their underlying motivations or needs.

FINDING RULE BREAKERS

Finding the right people to observe helps. People who follow directions perfectly and can't imagine a different course aren't much help. You learn more from the woman who takes a shortcut, who forces the product to do something the manual says it can't, who imagines what it might do if only . . . You learn from people who break the rules. We're often asked to do "futures" projects, a twist on the old "concept car" tradition. Companies frequently want to explore products or services or ways of doing things years into the future, and they often give us leeway to explore the product landscape. AT&T, for example, once

asked us to help it develop some new interaction strategies for business telecommunications. The objective was to help the company figure out how to make common phone tasks, such as three-way conference calls, more intuitive and easy to use.

So Jane and a few other team members paid a visit to one of our travel agents, a woman named Sally. While Sally's phone had a flash button for conference calls, she found it confusing. Like most of us, she had no interest in trying to decipher the complicated instruction booklet in order to figure out what "flash" meant or how to use it.

Camera in hand, Jane asked Sally to make a conference call. The travel agent went around the room, grabbing phones and placing them in a circle on her desk. One by one, she'd phone the people she wanted to talk to, putting them all on speakerphone so she could talk to them together. Sally loved the versatility of the system she'd cobbled together. If she needed to privately confer with one of the parties, she'd take him off the speakerphone and cup the handset to her ear. Or if she wanted just two of the parties to talk, she'd take them off speaker and hold the handsets next to one another.

Sally knew her system was out of the ordinary. "I know this is crazy," she confessed. "But this is what I do. It works for me." Jane understood why Sally had concocted this elaborate procedure. Sally was creating physical references. By giving each caller a separate phone and location on her desk, Sally could "see" and hear who was speaking at any point in the conference call. This was her mental model, a concrete routine she'd devised to make physical connections between people she couldn't see.

The extremes Sally went to in creating this physical and mental map dramatized the limitations of her phone. What if her phone provided more clues to let people "see" what's going on in a conference call? What if her phone could offer a natural, intuitive way of visualizing a conference call?

Sally's actions had a lot to say. It's a simple lesson that we often miss. Look at the Sallys of your business. The women and men who do

things a little differently. The people who might teach you something you didn't know.

YES, PEOPLE ARE HUMAN

There was a time when companies wrote off people like Sally as "stupid customers." A Big Three auto executive in the 1930s once lamented, "It's not that we build such bad cars; it's that they are such lousy customers."

The annals of business are chock-full of executives who didn't understand why people couldn't use their products correctly. For example, Iridium's ill-fated satellite phone system required customers to have what they called "user dexterity." Iridium expected callers to position themselves so that nothing blocked the line of sight between their phone and the orbiting satellites—an unfathomable geometry problem for some customers. Sooner or later, many of these companies lost market share or—like Iridium—went bust. Unless you wield monopoly power, such arrogance is generally not a good customer strategy.

Nor is treating customers like statistics. In these days of click-through demographics and detailed buying patterns, too many companies assume that the answers will be electronic, that everything worth knowing will be assigned scientifically determined percentages. But this assumes you know the right questions to ask. And it forgets that outrageous new products and ideas recognize that people are human.

Empathy is about finding and listening to the Sallys of the world. It's about rediscovering why you're actually in business, whom you're actually trying to serve, what needs you're trying to fulfill. Companies periodically need an empathy check. Often they fall into the trap of responding to what seems to be market needs: introducing new features simply because other companies are introducing new features.

Long-term market leaders especially face this pressure. A few years back, Kodak asked us to help investigate the future of digital imaging. As with many projects, there was a tension between quickly getting a product out and laying the groundwork for future strategies. One of the first things we did was to remind team members about the underlying emotions associated with collecting, sharing, and viewing images. Jane asked each of them to write a half-page essay about picture-related experiences they'd had in the past six months, anything from snapping family photos to sorting their album or mailing photos. Jane shared the personal insights with the rest of the team and published them in a little booklet, which we gave to each member.

Though only a few Kodak employees wrote essays, word spread to Kodak's advanced technology lab. Upon hearing one of these family photo-taking tales, one of the lab members said he realized for the first time how important photos can be to families. Clearly touched, the man had a newfound desire to make his project relevant to real people.

Inspirations like these don't always immediately spawn new products or services, but they can refresh and reinvigorate workers. Jane's work has had a similar effect on employees at IDEO. We don't always have the time or budget to extensively research all the human factors that may be at play. But because of Jane's example, we've all learned to consider more carefully the human component.

OBSERVATION EXERCISES

Observation and inspiration don't have to be formal. Each one of us can learn to be a better observer simply by taking stock of our environment. Try this simple exercise. The next time you visit another company—or even your own—watch and listen carefully. Is it clear where you're supposed to go? Does a receptionist greet you and ask you to go somewhere else? Is it a comfortable place to wait, or do you feel like you're at the dentist's office? Which parts of the process welcome you

like a special guest, and which parts leave you feeling like a drone in the hive of industry?

What about meetings? Who arrives late? Where do people sit? Who runs the show? Next time at the airport, think how you might reorganize the whole experience if you were running the show, everything from baggage check-in and finding your gate, to waiting for information on seat upgrades.

If you're like most of us, you'll probably find many of these human systems lacking. The office waiting room feels like a presentencing chamber, the meeting may be chaotic or overly hierarchical, and, let's face it, kindergarten classes are run more efficiently than some airline check-ins.

Pop psychology books say, "Listen to your inner child." But watch how the world works and you become hyperaware: Who treats you like a person—and not a transaction? Who values your time over their system—how many times have you spent extra time waiting in line to buy something because the store's computer register stumped the clerk or because phone calls got higher priority than waiting customers? Who is watching and thinking about what could be done better?

Somebody in the hospitality industry finally listened to their customers' needs. Notice how good business hotels now include an iron and ironing board? If someone had bothered to follow business travelers around, long ago they'd have noticed salespeople grimacing at wrinkled clothes, turning on hot showers to try to steam out the wrinkles, and, of course, calling down to housekeeping to get hold of one of the few irons in the hotel.

Athletes must stretch before competing. Similarly, our observation muscles could use a little stretching. Keep your "bug list" handy, and use it to turn minor failings into improvements for your organization. Even the best companies could better design some of what they do. FedEx, for example, is a great company. But why can't it give you a proper envelope with its monthly invoice? Everyone in the organization and many outside of it see the beautifully designed FedEx Letter envelope the company provides to customers for free, but the person who

pays the bill also sees the "Rube Goldberg does origami" envelope for the payment check. Is it really optimal to make customers tear, fold, and lick this quirky envelope? And if FedEx is "Absolutely, Positively" the best way to send business documents, why is the envelope prepaid for sending via the *U.S. mail*, when the FedEx person already comes to your office twice a day?

That proverbial blister in your shoe just might be tomorrow's innovation. Remember those ultrastiff "mountain climber" hiking boots twenty years ago? The salesperson would advise you with a straight face that you shouldn't take them on a hike until they were properly broken in, and routinely sold you moleskin to ease your pain. A good observer of that exchange would have sensed an opportunity in the making—a hiking boot that didn't draw blood.

LITTLE INNOVATIONS

Sometimes even the best innovations fail because of simple, preventable miscalculations. Nothing can make a design team more humble than seeing its "great" idea scuttled because the team didn't take the human factor into account. A few years back, Procter & Gamble asked us to help with its intensely clever new delivery system for Crest toothpaste. We worked with them a little on the stand-up tube, but IDEO's biggest challenge was the cap. You know how the threads on a regular toothpaste tube often get caked with old paste? Caps don't screw on easily after that, so they often get lost or simply aren't used. There's nothing less attractive than a half-used toothpaste tube with dried paste all over the tip. As a Procter & Gamble executive said in our first meeting, "We're going to save marriages with this product."

Why not a pop-on, pop-off cap? we wondered. Drop the screw threads entirely. Make a smooth cone around the top that cleans easily.

IDEO worked up some prototypes and then we began observing. And what we saw completely surprised us. People kept trying to *screw off* our pop-off cap, even after they realized it had no screws. Decades of screwing caps on and off—thousands of times for most people—had created an ingrained perception and habit: The cap on a toothpaste tube must screw on and off.

We first thought that people would get used to the pop-off cap. But we were fighting a very hard-to-break habit. So we worked up another solution. A hybrid. A one-twist cap. It still boasted a smooth, easy-to-clean cone at the top. But now there was a short thread below.

The people we observed preferred the hybrid, and Procter & Gamble, a company with some experience in consumer packaged goods, liked it too. So we went with *half* our innovation. And the public immediately embraced the Crest Neat Squeeze tube, buying up over $50 million worth of the product in the first year alone. Nine years and more than 1 *billion* units later, the package—with its one-twist cap—is still selling well with the same design.

Successful innovations recognize that people don't always do the "right" thing or make the necessary leaps to bridge the gap between familiar and genuinely new ideas. Widespread adoption often takes time. Astute observation is one way to shorten that cycle and make trade-offs that users will accept.

We once worked on a portable electronic device for use in hospitals, essentially an oversized mobile computer. When we realized it was going to weigh nearly twenty pounds, we knew we had a problem. Obviously, you could put it on a rolling cart, but in actual practice, we knew that nurses would be lugging these machines around the hospital. Since we were unable to further shrink the size or weight, a member of the team suggested that we shape it like a doctor's satchel from the 1930s, adding a generous handle to make it easy to carry with both hands if necessary. The handle had a bonus—it created a positive, visual icon that reverberated with the medical staff. Observations showed that nurses placed the machines at many different levels, sometimes even on shelves above their heads, so the team designed a

screen that could be tilted to view from nearly any position. Reaching out to the actual ways professionals use the product helped to make it popular among nurses and technicians and smoothed its adoption.

SEEING PRODUCTS IN MOTION

Sometimes you can see a new path toward innovation by starting with a relatively static object and thinking of it as a blur of motion. Take cell phones, for example. When Motorola first came out with its "brick" models in the 1980s, the phones were bulky, difficult to carry, and awkward to use. Advances in microchip technology and batteries gradually shrank phones. Now that phones have become fashion accessories, they're easy to carry, and we talk as we walk. They often fit in the palm of one's hand—or become invisible through the use of an earpiece.

We sometimes find it useful to think of products in terms of *verbs* rather than *nouns*—not cell *phones*, but cell *phoning*. We try to create products that are made for motion, like the Palm V and the Handspring Visor. See products as *verbs*—as animated devices that people integrate into their lives—and you'll become more attuned to how people use products, spaces, services—whatever you're trying to improve.

Several years ago, while working on the design of a cordless office phone for Dancall of Denmark, we watched people using phones around the office. We saw two things very clearly. First, people often struggle to stick their cordless phones into their shirt pocket. Second, colleagues are frequently irritated when someone's phone rings during a meeting. Our observations helped us to solve these problems. We tapered the phone in two directions to make it easier to slip into and out of a shirt pocket.

The irritation factor was trickier. We could include a ring that sounded more like a buzz, but not everybody likes his phone to buzz like a pager. Someone suggested a tiny flashing red light on the an-

tenna, visible out of the corner of your eye when the phone's in your shirt pocket. To complement that feature, we placed a large button on top of the phone that helped minimize the interruption. When a call came in, you could simply push the button which would answer the phone with a quick recorded message: "Please hold on, I'll be with you in a moment." You could then move quietly away from the meeting before taking the call. Small observations leading to small improvements, but if you keep that process up continuously, you can find yourself at the head of the pack.

Working on sports products has helped us to see things as verbs. Athletes are constantly in motion battling to gain precious seconds, and—not surprisingly—manufacturers must race to develop new products to help them keep ahead. Specialized Bicycle Components, an innovative firm that was among the first to mass-produce mountain bikes, came to us looking for new ideas on its ubiquitous water bottle. Though the company had helped popularize the bike water bottle and developed a huge market share, competitors were starting to nip at Specialized's heels.

To begin our search for new ideas, we watched bike racers and serious bikers using water bottles. This was particularly easy for us, since for years IDEO has had a pretty serious Wednesday night bike ride into the foothills above Stanford. Soon it was clear that there was a problem with water bottles, a problem similar to what we'd faced with the Danish phone. Eyes on the road, bikers often looked like relay racers missing the hand-off as they struggled to stick their bottles back into their bottle cage. That insight helped us to design an improved bottle with a tapered bottom—and a rubber friction ring to make the bottle easier to grip.

A second issue was trickier. Watching riders, we saw that taking a drink from a water bottle is a clumsy two-step dance—you have to draw the nozzle out with your teeth before you can squirt the water. This process is even more awkward when the bottle gets covered with dirt or mud from the trail.

The solution, as is so often the case, came from looking at products

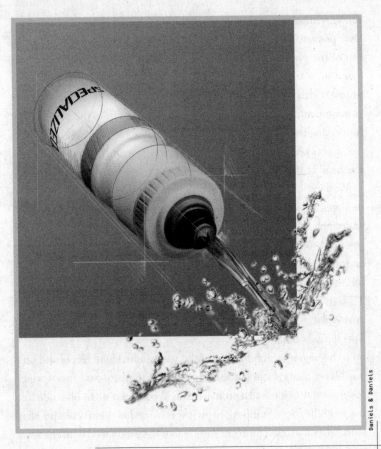

Daniels & Daniels

Watching mountain bikers inspired a water bottle that
keeps mud out of their mouths.

used for entirely different purposes. Our creative team members pe-
rused a range of innovations, including one of nature's clever designs,
the tricuspid heart valve. Three triangular patches of tissue open and
close the valves of the heart. Could a heart valve inspire a sports bot-
tle? we wondered. What if you sealed the top of the bottle with a rub-
ber septum that opens like a heart valve? Asking that question led the
team to design a simple self-sealing valve at the mouth of the bottle. A
rubber septum cut with an "X" let the water flow more rapidly than in

previous bottles. When you stopped squeezing, the septum would re-seal, not letting anything in. The bottle was always ready for a quick drink, but its contents would not spill. Or get overly dirty. All you'd need to do to get a drink was grab the bottle and squeeze—a simple one-step dance. And you'd never have to stick your mouth around the muddy spout.

I think the Racer's Edge water bottle is a great example of how even what seems like a static product can be improved if you see it in motion. Remember the idea of thinking in terms of verbs rather than nouns? One way to think of the improvements in the new Specialized bike bottle is through verbs—opens simpler, pours faster, and stores smoothly. Try it yourself. Watch your products and services on the run. You might find a problem or opportunity you didn't expect.

CROSS-POLLINATE

As you can see, it's critical to watch people in motion to fully under-stand a problem. It's a great way to look for ideas that leap traditional boundaries. We think this "cross-pollination" approach is a kind of alchemy of innovation. I'll be talking about it in depth a little later in this book, but I'd like to briefly give you a sense of how it often works at IDEO. Think back to that heavy medical device we talked about ear-lier. Before it was set to go to manufacturing, we spent some time ob-serving nurses. They'd control the electronic device with a computer stylus tethered to the machine. When a nurse was ready to change into a monitoring mode, she'd gather up the cord and snap the stylus back inside an enclosure. But sometimes after the nurses used the stylus, they'd look for somewhere convenient to temporarily store it, rather than put it back properly inside the enclosure.

We suddenly realized the stylus was like an old fountain pen looking for a holder. A team member wondered aloud whether the rubber sep-

tum we used in the Specialized water bottle might offer a better answer for temporarily storing the stylus as well. We had only a few days left before the final design was to be sent to manufacturing, so we tried a quick fix. The team punched a small hole in the screen's lower border and inserted a rubber septum. Sure enough, the septum made an ideal pen holder. So we ran with it. A minor improvement, you might think, but the kind of improvement that can only come about from watching people in action.

So many products and services have flaws that we start to assume that's just the way things are. But if people are frustrated or confused by a product or service you offer, eventually they may be wooed away when a better product or service becomes available. At the very least, rather than creating positive buzz about your product, they will say nothing at all or complain about it. That's why it's critical that you continue to innovate and improve. And the best way to do that is to watch people use your products.

Think about how frustrating and embarrassing the experience of buying a new or used car can be. For decades we—and the car companies—assumed there was no alternative. No one imagined you could buy a car without creeping anxiety that you were being cheated by fast-talking salespeople. When the Saturn came along with fixed pricing and a low-stress no-haggling approach, GM was obviously offering a completely different buying experience. People liked it. And with the advent of the Internet, the way cars are bought is changing once again.

The same is true of rental cars. Returning a rental car used to be an unpleasant and time-consuming task. Racing to get to the airport, you often worried whether a line at the rental counter would cause you to miss your plane. Now the process takes seconds. The mobile agent waves you down, enters your mileage into a handheld portable, and away you go. You might say that this innovation wasn't possible until portable computing devices reached a "wearable" size, but that's not completely true. The rental car companies could have set up portable stations in the parking lot decades ago. The fact that it was easier to do

as technology got smaller and more portable is only part of the story. What it really took was a change in *mind-set*. Rental companies had to recognize that they'd save money and build loyalty if they moved their services closer to their customers.

MAKING HEROES

Pay attention to how your customers might like to interact with your products or services, and a remarkable change takes place. You can do more than simply satisfy their immediate needs. You might actually make your customers feel like heroes.

Breakthrough products generally don't come to you in a flash of light. Usually, it's a gray day with no forecast of sun. Berkley Outdoor Products, for example, came to us with what seemed a waning ritual. Kids weren't fishing as much as they used to, and sales were down.

We went to stores and were struck with the frustrating complexity of fishing. Everything was sold separately: rods, reels, tackle, and line. Next, we went out to the fishing piers and suburban lakes to see the state of kid fishing, and found plenty of struggling dads.

In the old days, Dad could ask the old-timer behind the counter in the fishing store for help or information, but that won't work today at Wal-Mart. Dads often don't remember much about fishing and don't want to look stupid. So we decided to change gears. Make it all-in-one, playful, and toylike. Rod and reel incorporated in one piece, with all the essential gear in its own compartment—lures here and bobbers there. A complete set of instructions, including how to tie a hook on the line, laminated on the back of the plastic case. My favorite touch was the smallest, but it fit with the all-in-one concept: Even *bait* was included, in the form of a magical worm-shaped mixture Berkley calls Power Bait.

As a father, I especially like the way Berkley's new approach redesigned the fishing experience from Dad's point of view. Dad went from looking a little foolish in front of the kids to becoming the knowl-

By watching how dads struggled with their kids' fishing
gear, we came up with an all-in-one fishing kit.

edgeable hero. And this wonderful ritual pastime became easier to learn and pass along to the next generation.

If something's broken, it *can* be fixed. Just keep your eyes open for inspiration.

Everybody wants to be a hero.

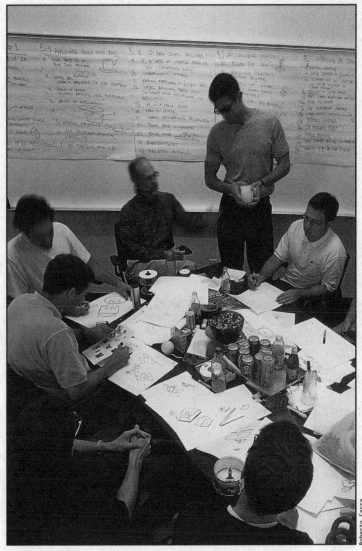

> The best way to get a good idea
> is to get a <u>lot</u> of ideas.
> —LINUS PAULING

The problem with brainstorming is that everyone *thinks* they already do it. When I raise the topic of brainstorming in presentations and conversations with business executives, I see some eyes start to glaze over with the smug "been there, done that" look of someone who has already answered "yes" to the question "Do you brainstorm?" In fact, more than 70 percent of the businesspeople in a recent Arthur Andersen survey say they use brainstorming in their organizations.

Many businesspeople treat brainstorming as a check box, a threshold variable, like "Can you ride a bicycle?" or "Do you know how to tie your shoes?" They overlook the possibility that brainstorming can be a skill, an art, more like playing the piano than tying your shoes. You're always learning and can get continuously better. You can become a brainstorming virtuoso. You can have heads of giant corporations and heads of government fly in to ask for your brainstorming help. (Don't laugh. I've seen it with my own eyes.)

So if you say you already do brainstorming in your organization, great, you're on your way. But I believe you can deliver *more* value, create *more* energy, and foster *more* innovation through *better* brainstorming. For one thing, you could brainstorm more often, weaving it into the cultural fabric of your organization. In that same Arthur Andersen survey of people who said they brainstorm, 76 percent admitted they brainstorm less than once a month. *Less than once a month.* I consider myself a movie fan, usually seeing thirty to forty films a year on the big screen (plus again as many on video), but if I dropped below one a month, I'd have to say I was a *former* fan. If you want to keep in shape, you have to exercise your brainstorming muscles more than once a month. So find a suitable space, order some supplies (and some chocolate chip cookies), get a good group together, and brainstorm up several dozen possible solutions to a problem that's bugging you right now.

Brainstorming is practically a religion at IDEO, one we practice

nearly every day. Though brainstorms themselves are often playful, brainstorming as a tool—as a *skill*—is taken quite seriously. And in a company without many rules, we have a very firm idea about what constitutes a brainstorm and how it should be organized. First, a brainstorm is not a regular meeting. It's not something you take notes at. You don't take turns speaking in any orderly way. It shouldn't consume a morning or an afternoon. Sixty minutes seems to be the optimum length, in our experience, though occasionally a brainstorm can productively stretch to an hour and a half. The level of physical and mental energy required for a brainstorm is hard to sustain much longer than that. Brainstorming sessions aren't presentations or opportunities for the boss to poll the troops for hot ideas. Nor should they feel like work. And brainstorming is most definitely *not* about spending thousands of dollars at some glamorous off-site location.

Brainstorming is the idea engine of IDEO's culture. It's an opportunity for teams to "blue sky" ideas early in a project or to solve a tricky problem that's cropped up later on. The more productive the group, the more it brainstorms regularly and effectively. We call the sessions "brainstormers," which to us sounds more like an active, engaging event. The buzz of a good brainstormer can infect a team with optimism and a sense of opportunity that can carry it through the darkest and most pressure-tinged stages of a project.

So what makes a brainstormer sing? Most people are familiar with the fundamentals—like sticking to one conversation at a time and building on the ideas of others—but it takes extra effort if you want a *great* brainstormer with valuable results.

SEVEN SECRETS FOR BETTER BRAINSTORMING

1. SHARPEN THE FOCUS

Good brainstormers start with a well-honed statement of the problem. This can be as simple as a question. Edgy is better than fuzzy. The

session will get off to a better start—and you can bring people back into the main topic more easily—if you have a well-articulated description of the problem at just the right level of specificity. A brainstormer without a clear problem statement is like a company without a clear strategy: You'll wander aimlessly and need a lot of extra luck or talent to succeed. If you find yourself leading a session that lacks direction, spend a few minutes developing a good problem statement. It will be time well spent. For example, "spill-proof coffee cup lids" would be a bad brainstorming topic because it's too *narrow* and already presumes you know the answer. Another approach, "bicycle cup holders," is too dry and product-focused. Maybe bicyclists shouldn't use *cups* at all, in which case they certainly don't need cup *holders*. A better, more open-ended topic would be "helping bike commuters to drink coffee without spilling it or burning their tongues." Go for something tangible that participants can sink their teeth into, without limiting the possible solutions.

We've also found that the best topic statements focus *outward* on a specific customer need or service enhancement rather than focusing *inward* on some organizational goal. For example, topics like "How can we regain market share from Company X?" or "How can we build a better search engine than Company Y?" may not lend themselves to innovations that will be embraced by the market. On the other hand, a series of more specific, customer-focused brainstorms on topics like "How can we accelerate the time-to-first-result for customers searching via dial-up modems" could uncover innovations that might ultimately yield the competitive edge you are looking for.

2. PLAYFUL RULES

Don't start to critique or debate ideas. It can sap the energy of the session pretty quickly. You need a way to turn aside critiques without turning off the critiquers completely. At IDEO many of our conference rooms have brainstorming rules stenciled in six-inch-high letters on the walls, for example, "Go for quantity," "Encourage wild ideas," or

"Be visual." Not willing to mark up your walls? Perhaps the facilitator could ring a bell when participants try to turn a brainstormer into a normal meeting. Or you could print the brainstorming rules on a giant Post-it or write them prominently on a whiteboard.

3. NUMBER YOUR IDEAS

Numbering each idea is pretty obvious, right? So obvious that it took us almost ten years to figure it out. Numbering the ideas that bubble up in a brainstorm helps in two ways. First, it's a tool to motivate the participants before and during the session ("Let's try to get a hundred ideas before we leave the room") or to gauge the fluency of a completed brainstorm. Second, it's a great way to jump back and forth from idea to idea without losing track of where you are. We've found that a hundred ideas per hour usually indicates a good, fluid brainstorming session. One hundred and fifty per hour is pushing the speed limit. Of course, it's *quality* that ultimately matters, but we've heeded Linus Pauling's advice to get a *lot* of ideas.

4. BUILD AND JUMP

Watch for chances to "build" and "jump." High-energy brainstormers tend to follow a series of steep "power" curves, in which momentum builds slowly, then intensely, then starts to plateau. The best facilitators can nurture an emerging conversation with a light touch in the first phase and know enough to let ideas flow during the steep part of the ideation curve. It's when energy fades on a line of discussion that the facilitator really earns his or her keep.

In the coffee-while-bicycling brainstorm example above, a good "building" suggestion to keep up momentum might be: "Shock absorbers are a great idea; now, what are some *other* ways to reduce spillage when the bicycle hits a bump?" By contrast, when discussion tapers off, a good "jump" transition statement might be something like this: "OK, let's switch gears and consider some totally 'hands-free' so-

lutions that allow the cyclist to keep both hands on the handlebars at all times. What might those solutions look like?"

Try building on an idea. Encourage another push or introduce a small variation. Or take a jump, either *back* to an earlier path you skipped by too quickly or *forward* to a completely new approach. Whatever you do, try to get into the next power curve and keep the energy up.

5. THE SPACE REMEMBERS

Great brainstorm leaders understand the power of spatial memory. Write the flow of ideas down in a medium visible to the whole group. There are many emerging digital technologies for group work, but we have had great success with extremely low-tech tools like Sharpie markers, giant Post-its for the walls, and rolls of old-fashioned butcher-shop paper on the tables. Brainstorming is an intensely group-oriented process, and the facilitator's rapid scribing is one of the focal points that hold the group together. We're not talking about taking personal meeting notes here, but capturing ideas so that the group can see their progression and return to those that seem worthy of more attention.

Cover virtually every wall and flat surface with paper before the session starts. That way you won't find yourself in the awkward position of having to erase ideas to make room for more. And you may find there's a certain synergy in physically moving around the room writing down and sketching the ideas. As you rapidly capture the team's ideas, make a mental note of ones that are worth coming back to during a build or a jump. When you return to the spot on the wall where that idea was captured, spatial memory will help people recapture the mind-set they had when the idea first emerged.

6. STRETCH YOUR MENTAL MUSCLES

People are busy. Time is short. Is it worthwhile to "burn" some time at the beginning of a brainstorm doing some form of group warm-up?

Maybe. But that "maybe" rapidly becomes a "yes" in certain circumstances:

- When the group has not worked together before
- When most of the group doesn't brainstorm frequently
- When the group seems distracted by pressing but unrelated issues

One type of warm-up we practice is a fast-paced word game simply to clear the mind (Zen practitioners call it "beginner's mind") and to get the team into a more outgoing mode. While these group exercises done in a circle may seem like you've been transported to your toddler's Gymboree class, they help ground you in the group and temporarily blot out worries about the latest operational glitch or the unread e-mails waiting back at your computer.

Another warm-up is to do some content-related homework. At IDEO we did a systematic study a few years ago to measure the value of different brainstorming warm-up techniques. We divided our pool of participants into three groups and had them brainstorm on applications of specific new technologies to the toy industry.

First we gave two of the groups homework. One group did some background reading and listened to an expert lecture on the technologies. Another group spent half an hour in a local toy store looking at the state of the art in all its variations. The third group came into the brainstorming session cold. Guess what? The group whose homework was a visit to a toy store significantly outperformed the other two, both in the quantity of ideas produced and in perceived quality. Not scientific enough to submit to the *New England Journal of Medicine,* but we were convinced. We now frequently bring "show-and-tell" to a brainstormer to help us visualize the wide variety of options and materials that could be applied to the session's topic.

One of the best brainstormers I ever attended at IDEO was an exploration of alternative wine beverage containers. Before the brainstormer, we covered a conference table with bottles, closures,

ncourage wild id

Steven Moeder

Printing brainstorming rules on the wall
can help keep sessions on track.

materials, and mechanisms ranging from the retro porcelain Grolsch beer bottle stopper to an elegant black Japanese sake flask. By the time we'd examined everything on the table, the session was already in full swing. We had hundreds of ideas before we were done, and a point of view on how to help our client innovate.

7. GET PHYSICAL

Good brainstorms are extremely visual. They include sketching, mind mapping, diagrams, and stick figures. You don't have to be an

artist to get your point across with a sketch or diagram. Leave your performance anxieties at the door and jump in with whatever visual tools you have available.

But the best brainstormers often get physical. We move beyond two dimensions and push for three. The first way we do this is to bring in everything but the kitchen sink (and we've brought the sink, too, when it was relevant). That means competitive products, elegant solutions from other fields, and promising technologies that could be applied to the problem. The second way we get physical is to have materials on hand to build crude models of a concept: blocks, foam core, tubing, duct tape, whatever might be useful. The third physical approach is "bodystorming," where we act out current behavior/usage patterns and see how they might be altered. We've bodystormed on products ranging from vending machines to car seats, where our skits and scenarios pointed to all kinds of opportunities for improvement.

THE BRAINSTORMER EFFECT

Hot brainstormers may generate a hundred or more ideas, ten of which may be solid leads. They can help put a team on course, and the rush of adrenaline can keep team members buzzing for days. There's a ripple effect. People talk after brainstormers, sharing wild or practical ideas that may have come out of a particularly vibrant session. A great brainstormer gives you a fantastic feeling of possibility, and an hour later you walk out of the room a little richer for the experience. I think that sense of spontaneous team combustion is why we've been able to find so many unusual solutions to seemingly intractable problems. For example, in designing the slim Dynabook laptop we discovered there wasn't enough room for traditional fasteners. A brainstormer generated the idea of "knitting" the two halves together—like a piano hinge—which sent us out looking for a strong, thin rod threaded on one end.

Daniels & Daniels

This surprising but effective use of a bicycle
spoke was inspired by a brainstorm.

We eventually reached a solution by getting our favorite Palo Alto bike shop to supply custom spokes to Dynabook that, amazingly enough, did the trick.

You can't quantify the value of letting people's minds run wild. At one brainstormer a few designers and engineers got the idea to motorize a swirling pencil—and made one on the spot. "We've struck gold!" somebody yelled out. We didn't follow up successfully on the whimsical toy. But a few years later, with or without our help, the Wigglewriter came to market and sold like hotcakes. A waste of time? Not in my book. It's the sort of incorrigible spirit of play that keeps team members in tune and on edge.

I'm convinced that if you participate regularly in brainstormers, it's like stretching exercises for your mind. You approach problems differently than when you're at your desk or strolling the halls. When a client rejects a course of action, the two or three team members working on that specific aspect of the project don't slink off to their desks to mope. They'll put up chairs to the nearest table and hold a mini-brainstormer, maybe just fifteen or thirty minutes to find a new track.

Brainstormers can do something else for your group. They offer the team members a chance to shine. It's friendly competition. Who doesn't want to demonstrate their cleverness and wittiness in front of their peers? And when a brainstormer thunders, it's a multiple win, a rising tide that lifts everyone up on a wave of excitement.

Some people, of course, are better at brainstorming than others. And truth be told, your performance *does* matter. Pipe up in brainstormers, bang out a quick prototype, or solve a tricky technical question, and you'll rise in the eyes of your peers. If you've got an open style of innovation—with brainstormers, prototyping, and problem solving—you'll naturally favor those who take action. It's true, "hot groups" aren't a place for shrinking violets. When a newcomer starts peppering a brainstormer with cool ideas, we take notice. You earn respect and authority in that week's brainstormer or on today's prototype team.

Unfortunately, we all occasionally fall victim to negativity. The coach who hammers too hard on the fumbling player, the teacher who corrects the flub and fails to praise the insight. Without a comfort zone, people don't take chances. If you're going to do a lot of brainstorming, there are several other pitfalls to watch out for.

SIX WAYS TO KILL A BRAINSTORMER

A brainstormer is a terrible thing to waste. It's easy to get it wrong. Here are six surefire ways to kill a brainstormer.

1. THE BOSS GETS TO SPEAK FIRST

If the boss gets first crack, then he's going to set the agenda and the boundaries, and your brainstormer is immediately limited. I know of one Silicon Valley boss who eagerly launched a brainstorm saying he was looking for some great new ideas. But unfortunately, he followed that enthusiasm with a room-silencing caveat: "Oh, and every new idea has to be patentable. And something we can manufacture." In that setting, nobody is going to suggest anything even remotely "wild." Try sending the boss out for coffee. Or doughnuts.

2. EVERYBODY GETS A TURN

I once sat in on a meeting a client *thought* was brainstorming. Sixteen people were packed into the room. We went clockwise around the table, and each person was given two minutes to speak. It was democratic. It was painful. It was pointless. It definitely wasn't a brainstormer.

3. EXPERTS ONLY PLEASE

Let's see, do I have a materials expert, an engineer, a software guru, and the V.P. of marketing? In brainstorming, don't be an "expert" snob. Bring in someone from manufacturing, who knows how to build things. Invite a customer service rep with lots of field experience. Find someone who reads a lot of science fiction. They may not have the "right" degrees, but they just might have the insight you need.

4. DO IT OFF-SITE

Brainstorming at ski lodges and beach resorts can be counterproductive. Do you want your team members to think that creativity and inspiration only happen at high altitude or within walking distance of an ocean? Don't get me wrong: Off-sites are fine. But remember, you want the buzz of creativity to blow through your offices as regularly as a breeze at the beach.

5. NO SILLY STUFF

Sports utility shopping carts. Velcro diapers. A privacy curtain to hide those embarrassing purchases. These are just a few of the wacky ideas hatched during our *Nightline* shopping cart brainstormer. I can't emphasize enough what these flights of fancy do for the team. They remind everybody that this isn't like work, that anything goes, and that we're going to have a lot of fun while we solve our problems.

6. WRITE DOWN EVERYTHING

I recently encountered a senior manager who's very good at a lot of things, but very bad at brainstorming. What's his main fault? He writes everything down, and I mean *everything*. He seldom contributes anything useful. We've seen this scenario before, especially among those who are new to brainstormers. Taking notes shifts your focus to the wrong side of your brain. It's like trying to dance and type on your laptop at the same time. Sketch all you want, doodle to your heart's delight. But don't act as if you're in History 101.

. . .

Everybody brainstorms. And everyone could get better at it, including IDEO—which was why we recently launched a research project to look for new approaches and variations. We believe it's worth the energy to try harder. A brainstorm can feel like it's just another meeting, or it can be a fun, invigorating experience that can take a project or a team to a new level.

Joe Watson

Teams are the heart of the IDEO method. It's no accident. We believe it's how innovation and much of business take place in the world. Quite simply, great projects are achieved by great teams.

Of course, we aren't talking about just any old teams. Though companies are naturally about groups and teamwork, too often these groups are simply the result of inertia. Years ago an "operations review committee" may have been formed, with broad and vaguely defined objectives. The group still lumbers on, meeting once a month, with no one feeling brave enough to suggest disbanding it. That sort of staid, unfocused group bears little or no resemblance to the hot groups we're talking about.

Hot project teams start with a clear goal and a serious deadline. The hot group knows that it might disband after the goal is reached and reform the next week to slay another dragon. How do you distinguish between dull and hot groups? It's the difference between administering a trust fund and making an MTV video. A hot group is infused with purpose and personality.

What's a bad group? A place where defending and preserving the group's status is a key goal. A place where everyone has a passion to retire. At IDEO we put that passion into the present. Our passion is about doing great projects. Today.

THE MYTH OF THE LONE GENIUS

If you distrust the power of teamwork, consider this fact. Even the most legendary individual inventor is often a team in disguise. In six scant years, for example, Thomas Edison generated an astounding four hundred patents, producing innovations in the telegraph, telephone,

phonograph, and lightbulb—*with the help of a fourteen-man team*. As Francis Jehl, Edison's longtime assistant, explained, "Edison is in reality a collective noun and means the work of many men." Even Michelangelo couldn't have painted the Sistine Chapel without the help of a gang of artisans. Prolific modern artist Keith Haring not only enlisted help on many of his own large projects but also started creating works of art in collaboration with literally hundreds of schoolchildren.

At IDEO we believe that the myth of the lone genius can actually hamper a company's efforts in innovation and creativity. After close encounters with dozens of real-life inventors, I have to report that most of them don't have a lot to teach us about applying the creative process to business. Too many of the inventors I have met suffer from a self-limiting form of paranoia. They want help with their inventions but aren't quite ready to reveal them. They aren't quite sure they can trust us with their precious secret and are worried that any potential partners will take advantage of them. So they return to the safety of their garages and basements and nothing ever happens.

The same is often true of the lone genius within a company. We've found that loners are so caught up in their idea that they are reluctant to let it go, much less allow it to be experimented with and improved upon. They think more like individuals than team members. And unfortunately, their projects suffer.

THE MAKINGS OF A HOT GROUP

History demonstrates that great projects and products are often the result of great teams. Products and projects as diverse as the Macintosh, the new Volkswagen Beetle, and the Lockheed Skunkworks were all spawned by charismatic teams. How do they do it? In a microcosm,

the IDEO *Nightline* shopping cart team—though they were only together for a week—displayed many of the characteristics of hot groups. First, they were totally dedicated to achieving the end result. No one doubted that shopping carts could use some improvement, and everyone was enthusiastic. Second, they faced down a slightly ridiculous deadline. When the hurdle is high, there's a tremendous sense of achievement in getting *anything* done by the deadline. Third, the group was irreverent and nonhierarchical. Despite the deadline, they joked and played around—like brainstorming up a sports utility shopping cart—to let off steam. Fourth, the team was well rounded and respectful of its diversity. Though the team was drawn from widely divergent disciplines, they had tremendous respect for their fellow members. You knew you were selected for your ability, not seniority or political skills. Fifth, they worked in an open, eclectic space optimal for flexibility, group work, and brainstorming. There were high ceilings with no internal walls, no sense that you often have in a typical corporate setting that "the company" wants it a certain way.

Finally, the group felt empowered to go get whatever else it needed. Hot teams connect to the outside world. They know that answers don't lie within. To design a better shopping cart, they sought out grocery stores and shoppers and industry experts. No one on the team thought for a second that a dozen people within IDEO had all the answers.

We've found that this model can help inspire all kinds of businesses, even those that at first glance don't seem conducive to teams.

HOT STUDIOS

At IDEO we've created a Hollywood-like system for quickly building teams around projects and disciplines. Successful film development companies have honed the art of quickly assembling teams—taking a

prototype (a script) and building around it the actors, directors, producers, and countless other individuals and contractors it takes to make a movie. In Hollywood there are no standing committees and no reliance on inertia. Every film production is an intense project where you field the best team you can, and race for the deadline. It's a model that works.

We've developed our own version of the studio system—and I think it's one of our biggest competitive edges. Design firms seldom exceed a hundred people. The reason is simple. When talent can't find enough places to blossom, it generally splits off in new directions.

David learned the importance of forming teams from his days at Stanford. It *mattered* to him who he worked with. But as the company matured, he realized he couldn't have a personal relationship with each person at IDEO; he couldn't be mindful of everyone's individual needs.

So as we approached the one-hundred-person mark at our Palo Alto offices in the mid-1990s, we took a chance and threw things open, designating some of our best people as studio leaders and giving them an opportunity to make their pitches to prospective members. No one at IDEO was going to be *assigned* to a studio.

Instead, at our Monday morning "all hands" meeting in Palo Alto, each leader described the type of work they favored, and what was exciting and challenging about their approach to product innovation. We reversed that horrible childhood sandlot routine where players wait to be selected. *IDEOers* got to pick their team leader, not vice versa. And you got to work *where* you wanted to: The studio heads each chose a location for their teams, among the various buildings in our Palo Alto campus.

Everyone gave David a "secret ballot" with their first and second studio choices, and amazingly, we were able to juggle the studio sizes so that everyone got their *first choice*. Each studio group ranged from ten to twenty people. Project teams, which operate within or across studios, depending on the nature of the work, could be as small as three people or as large as a dozen. After a couple of years, we opened it all

up again and gave everyone the chance to reshuffle studio membership if they wanted. It's worked. We've not only grown but also developed hot teams with very unique capabilities.

GENERATING YOUR OWN ENERGY

Creating a hot team is partly about mind-set. Hot teams don't necessarily have to start with the most exciting project. The right team and leader can create their own energy around the task at hand. My very first LP record (remember vinyl?) was the sound track of *Mary Poppins,* and I can still hear her saying that in every job that must be done, there is an element of *fun.* "You find the fun and—snap!—the job's a game." I am not seriously suggesting you quote Mary Poppins to the team, but the idea of making your own fun still applies.

People who work with kids already understand this principle, of course. Think of kindergarten (my son just graduated). The team—or class—spends nine long months slogging through some pretty dense material (like learning a new alphabet). A hundred and eighty days of school sounds like endless tedium. But the teacher (team leader) divides the time into "stories." Every month is something special. There's fall colors month, pioneer spirit month, dinosaur month, and so on. Instead of thirty-six undifferentiated weeks, every week or month has a meaning. Kindergarten is a series of projects, interspersed with games, challenges, and plenty of fun.

At IDEO we've developed lots of ways to build teamwork into even the most seemingly ordinary tasks, many of which involve competition and playful rewards. Dividing tasks into the acts of a play can give them more drama. Most accounting departments, for instance, appear at first glance to be unlikely places for hot teams to sprout. What's there to get excited about in all those rows of numbers? But think of

the intensity of the office when they're closing books for the fiscal year or quarter. What if the accounting team could tap into those rhythms and that energy more regularly? The accounting department is critical—and exciting—at those pivotal points. Build that enthusiasm into *every* workweek, and you've got the makings of a very hot team.

Finally, if you're hard-pressed to find inspiration, Tom Peters has some pretty good advice. He'd argue we can all benefit by thinking of our group or department as a professional service firm, or PSF, in its own right, with clients and projects and a seriously qualified professional staff. As Tom says, "Life in a PSF is as far a cry from Dilbert-Drones in DrearyVille (a.k.a. Departments) as one can imagine." No matter what you do, you can restructure the team's work practices to deliver better quality, faster results, or more fun. Peters cheers us all on to turn our work into "wow projects." Something you're proud of. Something that makes a difference. Something you can tell your kids about.

Easier said than done, you say. But focus on what's appealing in your work or project, and you'll be surprised at how teams will jell out of thin air. The attraction might be a challenging technical task. Or something with a positive social impact, say coming up with a medical product that can reduce pain or save lives. The possibility of public visibility can also inspire a team. But your project or team's contribution doesn't have to be trumpeted in the media. If it's bound to impress competitors, if it's something you feel proud to mention at a party, then it's hot.

THE PASSION FACTOR

What motivates a creative team? Passion plays a huge role. We can all learn from recent teams inspired by a cause or a search for excellence. Developers were dying to work on Linux because it was a chance to be

part of—and get some credit for—creating the world's best free operating system. Artists and technical staff at Industrial Light and Magic pride themselves on fashioning the world's best special effects in the latest hot movie. Paul Newman and Ben & Jerry's built an amazing amount of good will and customer loyalty out of connecting with worthy social causes.

Teams don't need to have higher causes to perform, but they do need tangible goals—a technical first, a challenging sales target, outshining a tough competitor. In short, something to shoot for and be proud of.

We believe the strongest teams take root when individuals are given the chance of picking what groups they work with and even occasionally what projects they work on. That way, passion fuels the fire. For example, we were recently asked to develop a kid's car seat. We asked for volunteers, and dozens of employees stepped forward. One of the dads who volunteered was so concerned about safety that he'd already bought *ten* different car seats for his three kids. He was picked, of course. To me, that's the kind of enthusiasm you need to boost a project to the next level.

Sports and recreation companies understand this kind of passion for the work. They live the lifestyle they're selling. That's why so many premier mountain climbers, kayakers, skiers, and windsurfers work for the Seattle-based sports retailer REI. For them, there's little dividing work and play. But your inspiration doesn't have to be physical, or even a conscious passion. It can start with a single emotional or intellectual chord. IDEO's car seat dad was helping to ensure the safety of his own children, as well as that of thousands of others. Not a bad reason to get out of bed in the morning.

How do inspired individuals become hot teams? Crazy deadlines and seemingly unreachable goals are often the sparks. In 1961 John F. Kennedy inspired a nation—and a lot of hot teams at NASA—by challenging them to put a man on the moon by the end of the decade. At the time, it seemed an impossible task. Yet America rose to the occa-

sion. In business, Steve Jobs is the master of setting the bar impossibly high, continuously challenging his teams to do something "insanely great." First he pampered and drove the original Macintosh team. Fifteen years later he resurrected Apple with the bold iMac when the company was almost written off as irrelevant (more than one critic prematurely penned Apple's epitaph) and in truth probably not far from extinction.

But opportunity can also inspire more humble business goals. Remember our *Nightline* challenge? We knew that if we could build a better shopping cart by Friday, we'd send a great message to millions of people. I think every hot team faces similarly challenging deadlines and goals. It's no secret. Pressure can help you get it out the door.

OPPORTUNITIES IN OBSTACLES

Adversity can prove the glue that binds your team together, making it emerge stronger and more tightly knit. That's what certainly happened with the hard case we designed and had manufactured for the Palm V. After making nearly ten thousand of the cases, our manufacturing liaison team discovered a slight quality flaw. Though the mistake was fixed on the assembly line, there seemed to be just two ways to deal with the existing inventory: throw out ten thousand cases, or take a chance that only a few dozen were defective and that customer returns would be limited. Fortunately, our team came up with a third alternative: After normal working hours, a dozen or so white-gloved IDEOers carefully pried open those ten thousand boxes and inspected each case, placing them into "reject" or "accept" piles. I did a shift, and it was a great experience. Rock music blared and we had a raucous time, cracking jokes until midnight. Team members would rotate through the different inspection jobs and then take a break for Pepsi and pizza (*without* the white gloves on).

It turned out that far fewer cases were defective than we feared, but it was the right thing to do for lots of reasons, not the least of which being that we would have been putting our reputation and relationship with Palm in jeopardy if we'd shipped flawed products. I think that little obstacle brought the team closer than ever before and reinforced its pride in the quality of products shipped.

Hot teams frequently take chances, though not always intentionally. A team in our Grand Rapids office was working on ways to improve airflow through an office space and give workers more control over their immediate surroundings. Suddenly, they hit on a new metaphor. Could the air vents in a car's dashboard provide some inspiration? Three team members drove to the nearest junkyard, paid the one-dollar cover charge, and set upon the first vehicle they came to. They pulled out their heavy metal—assorted saws, screwdrivers, and crowbars—and started hacking away. Suddenly, they heard the junkyard guy yelling that they were on the wrong side of the fence, just as they noticed the keys in the ignition. The team nearly panicked, as they all thought: "Jeez, I hope this isn't *his* car!" Fortunately, the car was just a junker for sale, and the team quickly set upon another carcass on the *right* side of the fence. They sliced out virtually an entire dashboard (at a total cost of $7.50) and got a leg up on their airflow question. The junkyard trip not only provided a breakthrough for the project but also became a memorable shared experience for the team.

ON A MISSION

Hot teams love a challenge. IDEO has designed and engineered countless products. But we've seldom manufactured and sold our own consumer electronics, a role we usually leave to our clients. Dennis Boyle's hot group decided it was time to challenge that status quo. In the late summer of 1999 as the team's intensive work on the Hand-

spring Visor neared completion, they began itching for a new challenge.

The Visor was an "open architecture" personal digital assistant, a Palm-like device designed to accept a wide variety of modules via its Springboard connector. By definition, its success, like the personal computer's success a generation before, would depend on the hardware and software of third-party, independent developers.

Dennis's team aspired to be one of those independent developers— to come up with the killer add-on. A product that would both be a hit in its own right and contribute to the overall success of the Visor. The IDEO team brainstormed up a few dozen alternatives and settled on the idea of making a small add-on digital camera for the Visor dubbed the "Eyemodule." This was a bold move, as the team would actually have to create and manufacture a highly technical product from scratch, something they'd never done. And to top it off, they'd never made a camera before.

By late September they'd assembled an exceptionally diverse in-house team, IDEO experts in everything from interaction design to high-volume manufacturing, along with some capable outside contributors. Boyle set what seemed an impossible goal of snapping digital images of Handspring's founders, Jeff Hawkins and Donna Dubinsky, by December 2.

The goal lit a fire under the team, and Boyle personally made sure they kept moving forward. Every few hours he circulated among the different disciplines to make certain everyone was communicating. When a designer complained that an electronic component didn't fit, Dennis skipped over to an electrical engineer and asked him to resize it. There was a constant battle to make sure the disciplines kept up with each other. More than once, Dennis made the interaction designer complete a design so the software programmer could continue writing the code and have a prototype ready for the circuit board due the following week.

Every week, Dennis held a team meeting, quickly shuttling through

A large group helped create this small digital
camera for the Handspring Visor.

the different disciplines, each subgroup covering only what was relevant to its task. At key points, outside consultants were conferenced in so they could query the team members.

Sixty days after assembling the team, they got the prototype working—a week before the deadline. Dubinsky was impressed but wondered how we could follow through on the myriad logistics and support issues. The team stepped up its deadlines—and made almost all of them. A dozen alpha units by the middle of January; three dozen by February, in time for Hawkins to demonstrate the Eyemodule at Demo 2000, the same place he'd held up the Palm Pilot four years earlier.

But the manufacturer didn't have the same sense of urgency, since we were still a small customer in their eyes. We experienced more than

our share of the usual manufacturing start-up issues, and the project team painstakingly fixed many of the early prototype boards by hand.

Throughout the process, the dedication and intensity of the hot team kept the project on track. When the purchasing agent went on vacation unexpectedly, IDEO team members stepped in, found the components, and ran them down to the factory themselves. When a bad batch of plastic parts didn't fit properly, half a dozen team members spent a couple of days at the factory, outfitting the Eyemodules with new cases.

As production ramped up, another problem arose. When the manufacturer couldn't automatically set the focus of each lens, team members rushed back onto the factory floor, focusing thousands of lenses by hand. IDEO learned a new technique it was able to pass on to the manufacturer, who ultimately was able to automate the process.

Despite all the obstacles, by mid-April the Eyemodule was shipping. Incredibly, we already had a customer service call center up and running in Salt Lake City and developers making software for the unit. Manufacturing has since transferred to a plant in Malaysia, where it is now sailing along with hardly a problem. Today the Eyemodule is the best-selling add-on for the Handspring Visor, which is challenging the Palm line in the hugely popular handheld market.

I think that's pretty amazing: six and a half months from founding a team to getting a highly technical hardware product on the market. Not bad for a group that was learning some aspects of the process as it went along.

TEAM PROBLEM SOLVING

Teams find ways to leap barriers. Take the experience one of our team members had at a Silicon Valley company. Upper management wasn't giving his team the financial support it needed. Several subcon-

tractors refused to do any more work on the project until they were paid. So the team leader drove to the bank, withdrew $18,000 from his personal account, and paid them. Then he filed an expense report for $18,001.54 (he billed for the mileage to the bank). And the project got done. That kind of grit and determination, buoyed by a willingness to bend the rules when you have to, is one of the reasons hot teams are so effective.

Indeed, it's often the last-minute snafu that brings a team together. One IDEO team working on a project to reduce noise in office environments arrived at the conference room where they were scheduled to give a demonstration the following morning, only to discover that the room was acoustically isolated—virtually soundproof. There was no background noise to cancel out. It was the opposite of the hectic, noisy office environments that the team was developing solutions for, and would be disastrous for a demonstration. This unexpected hurdle set the team members to drowning their sorrows over beer at a local bar, where they realized they could barely hear one another. Suddenly, a lightbulb went on! One of the members ran out to buy a tape recorder. They taped the bar noise, and the next morning used that background noise in their demonstration, which was a hit.

TEAMS IN CLOSE QUARTERS

We've all had that heady feeling at one time or another. I'll bet you have been part of a hot group. One of my personal hot-group experiences took place while working as a management consultant in southeast Asia. The focus of the project was a bit fuzzy. Singapore Airlines hired us to do a review of their already top-notch engineering and maintenance operations. Our three-man project team divided this complex problem into manageable parts. Each senior consultant tackled an area—operations strategy, inventory management, and maintenance scheduling.

The team did everything together for more than six months. We worked all day together, lived in the same hotel, ate practically every meal together, worked out at the same health club—even drove together to Malaysia twice a month to get our passports stamped. Squeezed into one small office, we had the chance to bounce ideas off of each other several times a day. And we often worked late into the night helping each other analyze and interpret the mountains of data.

Ultimately, we found the proverbial needle in the haystack, a small but costly flaw in this otherwise sterling airline. A miscommunication between the maintenance and marketing groups had resulted in a miscalculation about how much downtime a new fleet of aircraft needed. We showed how they could dramatically (and safely) reduce that downtime, increasing the number of hours per week the fleet was available for revenue flying. All those late nights the team spent poring over the data together had finally paid off, saving the airline tens of millions of dollars. Without the intense teamwork, I doubt if we could have made such a breakthrough.

Close quarters can charge the dynamics of a hot group. One of our hottest studio teams meets for its biweekly lunch in a ridiculously cramped place, and I doubt the intensity would be quite the same if the room weren't so crowded. Too much square footage, like too large a budget, can dissipate energy and discourage a more immediate and emotional connection between team members. One IDEOer had been a principal in a start-up and toiled with his two partners in a tiny, windowless space for a year and a half. Though on paper it sounds pretty bleak, in fact there was a great energy to the team. Later they got funding and their office space leaped from a thousand square feet to ten thousand. Suddenly, they had private, noncontiguous offices. Communication deteriorated and relationships soured almost overnight. The company eventually grew to fifteen people, but they'd lost momentum and ultimately never achieved success.

Of course, cramming people into tight places isn't the only way to

build hot teams. An impromptu setting can help spark team chemistry. When the Danish hearing-aid company Oticon discovered that spontaneous interactions between employees located on different floors were taking place on stairwells, the firm wisely broadened the stairs to encourage the multidisciplinary interchange.

The hottest teams flower like a budding relationship—somebody literally moves their clothes into your closet. While on a furniture project, for example, a handful of IDEOers worked on-site at the client's facilities south of San Francisco. They carved out their own space in the corner of the factory and worked and lived with their prototypes and counterparts. Their joint effort had the fervor of a mission. I'm not surprised that four months later they came up with a series of nifty new products for the client, including a rolling T-shaped office divider, capped with a whiteboard that's handy for small impromptu meetings or writing quick notes.

THE RIGHT MIX

There's an art to putting teams together. Personality plays a part. Sometimes it makes sense to stick an extrovert with an otherwise contemplative team, and vice versa. Sameness is not the goal. There are many facets to team dynamics. Where a team member sits, for example, can make a difference. Putting a particular engineer in the right intersection can mean the difference between success and failure. It's a puzzle we juggle all the time, as teams for projects form and disband. IDEO studio head Dave Blakely constantly finds himself creating little neighborhood teams of engineers culled from across the spectrum—from mechanical to electrical, software to manufacturing. The idea of establishing "neighborhoods" is to get the team as close to the hardware they're working on as possible—and make it easy to hold quick, spontaneous local meetings.

If you're working on a series of electromechanical products, you don't want to stick all the electrical engineers in one building and the mechanical engineers in another. Colocating the people who will need to regularly interact with one another recognizes the primacy of project teams and announces that the team is the most important unit in the company.

ABOLISH "THEY"

One of the things that struck me the most about my first days at IDEO—other than the fact that I had no desk or office—was that try as I might, I couldn't find a "they." At my old job as a consultant for a large international firm, there was always a spoken and unspoken "they." If the coffeemaker was broken, "they" should fix it. If you were entangled in bureaucracy, "they" were the problem. "They" do not innovate. "They" leave less elbowroom for you and other individuals to solve problems from the ground up.

Technology has in some ways exacerbated this problem. I'm confounded at how many companies bring "they" in reach of your every keystroke. I think those large corporations who snoop on their workers' e-mail are nuts—and not just for ethical reasons. If you practice monitoring or surveillance, your employees may end up being *less* productive and certainly less creative. But "they" companies can't help themselves. They're control freaks—even when they imagine they're loosening their grip.

A friend at a major transportation company sent me her employer's memo about its newly enlightened approach to office attire. The company had finally decided it was time to loosen up. "They" suggested there had been a lot of confusion and anxiety about what to wear on "casual Fridays." So "they" formed a task force. "They" issued a memo saying, among other things, that if you're unsure about whether something is appropriate or not, it's probably not. The memo went on to list

items of clothing that were definitely not appropriate, including jeans, shorts, and open-toed shoes—basically what you could expect to see at IDEO every day.

TEAMS AND TRUST

So what makes a successful team?

A few years ago I was on the same speaking agenda with Lou Holtz, the veteran football coach at the University of Notre Dame. Although I'm generally leery of sports/business metaphors, the coach said something that day that stuck with me. He said that the team changes every year, but each team member's three implicit questions for him remain the same: Do you care about me? Can I trust you? Are you committed to the success of our team?

The best leaders answer those questions with *actions* instead of words, and the best teammates leave no doubt about where they stand.

When my brother David joined Boeing in the early 1970s, the company assigned him to a mechanical engineering group. When he pointed out the mistake—his degree was in *electrical* engineering—no one bothered to reassign him. Instead, they stuck him in a massive windowless room with two hundred other (mechanical) engineers. Adding insult to injury was the strict work schedule policy. The workday began at 8:00 A.M. Arrive early and you couldn't get in. Show up twenty minutes late and they treated you like a grade school truant, taking your name down at the gate, giving you a note for your supervisor to sign. Boeing was, and is, a great company, but in that era, it was answering a resounding "no" to Lou Holtz's question about caring about its team members.

There was a time when mutual mistrust was the order of the day. Companies used to have office supply managers. More often than not, the prevailing attitude was that workers cared so little for their em-

ployer that given the chance they would rampantly steal office supplies. It is an attitude that makes no sense when team members think of the company as their own. That's why so many companies today are generous with tools. We try to make it easy for our teams to buy or use whatever supplies they might need. Nor can you go wrong by keeping refrigerators well-stocked with drinks and having a wide variety of food in ready supply, as many start-ups do. It's good hospitality and good karma. If your company is generous, workers are more likely to be generous in kind.

There's another reason to eliminate the "Office of They." Suddenly, every little snafu becomes an opportunity to exercise your collective problem-solving abilities. A take-charge attitude takes over. More than once, we've done a little plumbing or performed a little "first aid" on our buildings when they needed attention and the property management group wasn't handy. One reason an eclectic and self-determined approach to offices and worker neighborhoods is so effective is that there's no "they." There's just a group of people responsible for our team and our work space.

Strive to create an abundance mentality. Pick projects and clients that excite you—or better yet, you pick each other. You'll find yourself facing fewer turf battles. When our Boston office lands a client close to our Chicago office, it's not a cause for friction. There's a sense that every office will find plenty of work and that things will balance out in the end.

There's something terribly liberating about applying an abundance mentality to your ideas, thinking, and work. There's a Zen-like force here at play: The less you strive to control ideas and insist on credit for those that are yours, the more good ideas you're likely to have—and see implemented. It's why we believe in brainstorming. Good things happen when you willingly take your best ideas and pour them into the mix.

WHY TEAMS BEAT INDIVIDUALS

The myth that creativity flourishes primarily in solitude is an established one. We've all read over and over again how novelists and screenwriters writing in the seclusion of remote cottages or bungalows churn out best-sellers and blockbusters. But is that the only way to foster innovation? And is it a viable model for business?

Consider Academy Award–winning screenwriter Ron Bass. He has a team of six writers and researchers. They work up ideas for scenes. They research all kinds of details and trivia. They brainstorm. The team dives into two- or three-day outlining marathons, even speaking the dialogue out loud. Finally, the team blocks out the basics of scenes and dialogue, "early prototypes" in the IDEO vernacular. Then Bass sketches a scene in pencil, his assistant types it, faxes it to the team, and they call or fax back their responses. Generally, if half the team hates something, it gets changed.

Other screenwriters complain that Bass & Co. is a "factory"—and maybe they're right. It's an idea factory. Bass has written or cowritten sixteen feature films, including *Rain Man, Waiting to Exhale, My Best Friend's Wedding, Stepmom,* and *Snow Falling on Cedars.* And his films have earned over $2 billion. Clearly, he's doing something right.

Quite simply, Bass's team approach works. Team members scour novels for possible adaptations. A couple of them handle the stream of "notes" from studios demanding changes in plot or characters. Like Edison's Menlo Park laboratory, Bass and his team manage to constantly juggle four or five projects at once. Bass now turns out more scripts than ever before, and feels his team has improved the quality of his work. Momentum and synergy may be his team's greatest weapons.

Variety can inspire teams. By working on more than one project at a time, team members are often able to see the connections between different projects and emerge energized instead of exhausted. Bass can't fathom writers who complete a script and then shut down their creative juices and go on vacation. He and his team are so engaged that it "all continues: comedy, drama, buddy film, caper," he says. "It's all

one story, really, the story of who we are and how we relate and how we get it wrong." How does Bass keep his creativity machine running? As soon as he writes a script's final lines—"Fade to black"—he pens the first scene of his next movie. It's a wonderful strategy for any team—to end one project by beginning another. That way, when you do take a break, you already have an opening, a beginning to work from, rather than starting from scratch.

HOT TEAMS PERFORM

The best teams are like a snowball rolling down a hill, gaining in mass and momentum as they go. Very quickly, just a few snowflakes build into an avalanche of energy and enthusiasm.

How do you start building such a team? First, forget what you learned in school. Believe that your team members will be an outrageous success before their first day of work. That's what Benjamin Zander, conductor of the Boston Philharmonic, does. The celebrated music teacher has a unique strategy for getting the best out of his students. On the first day of class, he tells his first year students that they all get an A. There's one condition. Their first assignment is to write a letter to him—dated on the last day of class—explaining why they deserved the grade.

Zander, who wrote a book called *The Art of Possibility,* believes there are several beneficial aspects of this dramatic role reversal. First, it's a great confidence builder for his students. Second, it eliminates the often counterproductive sniping that people frequently engage in when they think only a few will win (one reason we believe bell curve grading is so flawed). But most important, students invariably knock themselves out for that A. Zander believes that they do more to earn their own personal A than they would ever do for the traditional A given by a teacher.

We do much the same at IDEO. It starts with the hiring process. The typical job candidate will be interviewed by more than a dozen of us before getting the thumbs-up. The process takes time, but it's worth it in the end. You don't get hired at IDEO unless you wow a bunch of us. Those that make it receive a tremendous boost, knowing that a lot of other accomplished people think they're talented and capable.

We believe in giving individuals an opportunity to perform. Everyone starts out roughly equal and then is given lots of chances to mess up—and to shine.

Newcomers that flourish in our environment are often offered a key role in a new project, or even an opportunity to manage a project. Age and experience aren't factors. You actually get to pick two of the three people who will review your work, and IDEOers invariably pick team members. And since we live for projects, there's an opportunity to spread the work around.

HOT TEAMS GIVE

We frequently buy the products we design—Palm Vs and Handsprings are ubiquitous, as is Crest Neat Squeeze in the bathrooms. Over the years, employees have found their virtual Christmas stockings stuffed with the likes of AT&T phones, Smith ski goggles, Nike sunglasses, and Polaroid I-Zone cameras. It's not practical to buy everything we help engineer or design, but we do our best.

Try giving lots of gifts and awards. But not out of routine. The creation of the gift itself can reinforce your team spirit. Even the company's holiday-season card can be an opportunity. We pride ourselves on our unusual creations, and one year even mailed out cards that contained tiny live trees. We used a space-age bonded plastic substrata, originally developed by NASA, to keep the trees alive. IDEOers pitched in after work to place three pinkie-sized redwood seedlings in each of the six hundred holiday cards.

End of Project Awards celebrate a team
member's superhuman efforts.

But gifts given by fellow team members build the strongest bonds.
End-of-project awards are a great way to punctuate team accomplish-
ments, celebrate contributions, and, yes, poke fun. What's important is
that everyone play some part in creating the award. The best end-of-
project awards make light of a blunder or cleverly capture the heart of
an individual's achievement. They can be strikingly simple. Early in his
career, for instance, one of our studio heads sent a reversed drawing to
the tooler; his award was a mirror on a plaque. Team members thought
another engineer may have overspecified the metal enclosure on a

large medical device. His "bulletproof" award was a hunk of sheet metal laced with bullet casings. Then there was the engineer working on a photocopier that initially had some paper-handling problems—which earned him an end-of-project award that included pieces of paper folded like fans. Or the team leader who spent weeks on a farm in Michigan improving veterinarian tools. His end-of-project award was a dartboard of a cow's derriere.

Sometimes, of course, awards just celebrate achievement. For four days, Andre Yousefi from our shop barely slept while trying to complete the *Nightline* shopping cart. His fellow shop workers rewarded him with a big Superman doll holding a small cart. And an engineer who helped hurdle a competitor's seemingly insurmountable sheath of patents was given a hatchet symbolically hacking through five hundred pages of printed patents. When an especially large joint team from two IDEO offices designed an information appliance for Italy-based Merloni, everyone gathered together for a team photo, holding up the dozens of early prototypes. The few members who missed the photo shoot were digitally added later.

But don't reserve fun for the end of a project. Some of our teams have designed custom T-shirts for their leader's birthday, and plenty of people have returned from trips to find their office spaces "redecorated" by their colleagues.

HOT TEAMS SHARE

How can teams stay motivated and abreast of the latest developments? One way is to launch a clubhouse, a regular weekly or monthly meeting that functions a lot like a writers' group. The point is not just to share the latest advances and technologies. Clubhouses help get us out of our routine, pull us up above the fray, and remind us of our pas-

sion for our chosen fields. They build camaraderie in a supportive, competitive environment and offer a semistructured method of maintaining a spontaneous, constantly evolving cutting edge. Clubhouses are part of the fabric of Silicon Valley, a culture known for user groups and home to such legendary innovation-crazy groups as the Homebrew Computer Club, which helped spark the personal computer industry.

A clubhouse can give you a sense of belonging to something larger than yourself. That's part of the reason we believe in the importance of giving tours of our workplace. If the place where you spend more than half your waking hours is considered valuable and even exciting to outsiders, think of what this does to the psyche and spirit of your own team members.

Tours can build pride and inspiration. But they're just one way to open your building and staff to the rest of the world. Over the years clients have occasionally asked us at IDEO for more than traditional product development and design. A number of companies have sent teams of engineers and designers to learn from the IDEO experience. In some cases, they've spent months at a time in our Palo Alto facilities, renting homes or apartments nearby, working on specific projects while trying to glean as much as possible about our methods. Others visit for a few weeks and then return periodically. They pick up on work practices that reinforce the culture. They try to absorb the nuances of our open-ended quick prototyping process and to observe the way it enhances the bonds between team members from different disciplines.

Several years ago a number of our Fortune 500 clients asked us to package a two-day version of this IDEO experience. That was the genesis of what we now call IDEO University or simply IDEO U. And just as *Nightline* chronicled our Deep Dive, an intensely compressed exercise in our practices and methods, IDEO U. has become an opportunity for outsiders to practice our methods firsthand. For us, it's a two-way street. We get to continuously test and reformulate our methods, learning what works or doesn't work—for different industries and

teams. And our IDEO U. teams serve as ambassadors, reporting back to the rest of IDEO in meetings and informal talks.

By expanding the horizon of a hot group, and keeping it ever evolving, you can help keep your team on the cutting edge of innovation.

TEAM MORALE

When people feel special, they'll perform beyond your wildest dreams. It's why we throw blowout end-of-year bashes—to celebrate who and what we are. We once managed to take temporary possession of Moffet Field's blimp hangar. Amid the airplanes, a zydeco band jammed, while we rollerbladed through an indoor space the size of seven football fields.

The point isn't how much money you spend but what kind of experience you can create. The main attraction at one of our recent end-of-year bashes was a gingerbread house design fest. We put dozens of prefab houses on tables and laid out all the supplies you could imagine—baker's tubes of confectioners' sugar, gumdrops galore, Hershey's kisses, red hots, and candy canes. The idea was so popular that more of us were hunkered down making baroque gingerbread houses than kicking up our heels on the dance floor.

Another great way to make people feel special is to let them play hooky. Sometimes the best inspiration for a team can be the Zen-like act of not doing any work at all. Like taking the afternoon off with your team to see a baseball game or watch a hot new movie (we once had a self-appointed "B team" that watched bad science-fiction films and other B movies on a regular basis). Since we've played a role in the making of a number of films, from *The Abyss* to *Free Willy*, we've taken in a matinee more than once. When *The Abyss* was released, we took over an entire local theater for a private screening.

You can't worry about how much time or money the fun costs. You'll

get back whatever time the team lost in the next week and quickly move into the black. After all, morale isn't something that can be measured or planned. Encourage unplanned breaks during the day. At IDEO a break might be a bike ride, or blowing off tension by shooting soft foam Finger Blaster rockets at one another, or, if it's winter and you're outside the Boston office, playing hockey.

The best diversions are combined with a prank. For one office retreat, Dennis Boyle's studio sailed around San Francisco Bay. The group knew they'd be sailing by our San Francisco office at Pier 28, so they put together a giant slingshot fashioned out of a big funnel and bungee cords soon dubbed the "funnelator"—and launched a merciless water balloon attack.

Field trips can be a great form of play. Try scooting off to see something that excites or inspires your team. We regularly take in the latest exhibit at the San Jose Tech Museum. Lately, we've begun taking field trips to other companies. Several IDEOers were recently invited to a daylong visit to Industrial Light and Magic and George Lucas's incredible Skywalker Ranch. It was great fun, and who knows, we might just work together someday.

The best field trips merge fun and projects. Before joining IDEO, David Haygood hosted a wild field trip for Hexcel, the maker of high tech composite materials. Teams from Boston and California converged for an inspiring week in Arizona. In the morning they had what they called "oyster sessions"—throwing a grain of sand into the group in the hope that it might turn into a pearl. The sinewy group scheduled a competitive event each afternoon—a 10K run or a long bike ride. They met for dinner and conducted a second work session in the evening. The exercise seemed to take the pressure off and give the morning brainstorms time to jell. Several good product concepts evolved out of the mix.

HOT TEAMS MEET

Taking field trips, playing hooky, and holding edgy off-sites can help keep a team motivated. But when you can take the energy that most companies only manage to drum up in an off-site meeting, and find it in your routine work, then you're onto something huge. The secret, I think, is to scuttle rules, provide good food, and encourage lots of play. Here's a glimpse of one such group at IDEO.

Two Thursdays a month, Dennis Boyle's thirty-five-member studio meets "Under the Wing" for lunch, featuring pasta, enchiladas, and salads. The seats around the conference table disappear nearly as fast as the food. Latecomers drag chairs into tiny spaces, and the stragglers squeeze through the crowd to grab a plate and a place to stand.

The atmosphere is a little bit like the kitchen at a bustling party. There's not really enough room for everybody, and that's part of the attraction. I try to make the Wing lunches whenever I can. It doesn't matter what's on the agenda because, strictly speaking, there *is* no agenda. This is playtime, a chance for show-and-tell, and, of course, a few jokes and harmless ribbing. In principle, Boyle runs the show, but his real job is just to keep the ideas flowing.

The studio uses bimonthly lunches to share insights and progress on projects. At one meeting, for instance, one of our senior designers stood up to talk about the sleek Webslate his team had just created, a "concept car" demonstrating the handheld possibilities of the new Transmeta chip. "It's eye-catching, it's for Web surfing, not so thin that it can't happen," he said, proudly passing around drawings. "The Palm concept, maybe taken to another level." Two weeks later Transmeta and Linus Torvalds unveiled the chip and Webslate at a major press conference.

A veteran IDEO team leader spoke enthusiastically about how voice recognition and pen technology were quickly coming together for handheld devices. He couldn't get into the details, but his team was building a "wearable" computer for a very large computer company. Excited by the possibilities, he talked about Microsoft's "ClearType"

screen technology and how close it was getting to the quality of type-script on paper.

Another team member showed off a bug-shaped information appliance for Generation Xers. Someone else talked about a defeatured, inexpensive cell phone for kids and the elderly, with a streamlined interface and an alternative to expensive monthly service charges. That led, naturally, to a discussion of the latest ear clip Ericsson phone and the new minuscule Nokia model, which someone had just purchased for seven hundred dollars and had already begun to disassemble.

And there was still room for a birthday celebration and some hilarious recounting of New Year's escapades, as well as nominations for the weirdest gift—an egg cuber—and the latest candidate for our collection of amazing edibles—hologram candy. Finally, the team discussed renting a go-cart track in the spring to celebrate the completion of a major project.

We've all endured countless dull meetings over the years, but most of our studios have the team meeting down pat. What makes it work is the enthusiasm, the show-and-tell, the humor, the irreverence.

HOT TEAMS LOOK THE PART

Whatever your current T-shirt budget is, double it. T-shirts cost practically nothing yet offer the return of team solidarity. By definition they're about teams. IDEO veterans have a couple dozen T-shirts. The best are edgy, inspired by projects. Once hooked, you'll find yourself unable to resist making more T-shirts than you ever imagined. T-shirts are like quick and dirty prototypes, and if your team or company has a number of projects going on, every few months you'll have something new to wear. Even if your team doesn't boast a hot profile, it can still sport attitude. Our accounting and marketing departments made a fleece vest that proudly shouted, "IDEO Overhead."

T-shirts are metaphors in motion. They have an urgency about

them. But if T-shirts don't work for your company, don't fret. Lots of us wear our passion in sweatshirts, bicycle messenger bags, or baseball caps. The original Macintosh team members were allowed to etch their signatures into the machine's injection molding tools, thus leaving their literal imprimatur inside the product's caseworks. Programmers working on cool software are notorious for leaving their names and nicknames buried within the source code. Let irreverence sprout when the spirit moves you. Five early Yahooans lobbied for the right to get their cars "shrink-wrapped" with a plastic purple and yellow Yahoo sheet. Grassroots team efforts like these are infectious. It's no surprise that some of the world's strongest brands started with a hot team.

HOT TEAMS NEED CHARACTERS

Remember that innovation stalls with the status quo. Celebrate your team members' differences. Many years ago one of the most brilliant engineers at Atari was, well, eccentric. He lost his apartment, so he literally moved into the office, taking residence in a makeshift loft that fellow team members built for him. Incredibly, the only entrance was through a conference room. He'd crank away all night, and you'd arrive in the morning to find all the amazing things he'd accomplished. But you couldn't talk to him till the afternoon when he clambered down from his loft. If you think this sounds bizarre, sometimes that's how innovation happens. Amazon's first and most prolific reviewer was practically entombed in books. Every wall of his office was crammed with volumes stretching up to the ceiling. His sleeping bag was tucked under his makeshift desk. He rarely went home.

Hot groups tend to have at least one or two certifiable weirdos. Look around your group or team to get a sense of the different roles people play. Most teams have everything from dramatic leads to character actors, writers, and, yes, directors. Characters round out your company,

and we all need a few individuals we're not sure what to do with. We've been lucky to find many of these talented types at IDEO.

EIGHT CRAZY CHARACTERS FOR HOT GROUPS

The Visionary

We all occasionally make the mistake of not rising above the fracas long enough to ponder what's next. At IDEO we have an elder statesman who inspires affection and enthusiasm among his flock. Over the years he's recruited key team members with divergent talents and personalities. He coined the term "interaction design," anticipated the importance of computer-aided design, and is the one most likely to figure out what's next.

The Troubleshooter

It may be politically incorrect to say it, but business bogs down with excessive politeness. There are plenty of times when we all could use someone who doesn't suffer fools gladly and cuts to the chase. At IDEO that's the first person David hired, IDEO's chief problem solver and fix-it man. His clear thinking and quick actions can right a drifting brainstormer or put a wandering project back on course. Devoid of political fears, our troubleshooter has occasionally rubbed a worker or client the wrong way, but his clarity of thought and expression has saved us dozens of times.

The Iconoclast

Every great organization needs a counterpoint, someone who challenges the status quo. We've got just that person, someone who doesn't buy into all of our ideas about teamwork and group innovation. Frankly, we're happy to have his contrarian view. Our iconoclast is a Silicon Valley design legend who broke the mold. He's got an uncanny ability to spot key business trends and strategic alliances for IDEO. And he's seldom wrong.

The Pulse Taker

Every company needs a heart. The head of our human-factors group makes sure we make a personal connection. One day she might tag behind a Japanese salesman to see how he really uses a portable computer. Another day she might invent archetypal characters who form the movielike narrative of how people might use a product or service. She inspires IDEO designers and engineers by unlocking the all-too-often-overlooked insights and patterns of exceptional and ordinary people.

The Craftsman

Just as we believe multidisciplinary teams are critical, we also believe that it's invaluable to bring certain talents and technologies in-house. In the age of the Internet and virtual corporations it's easy to think that expertise is just a click away. But you can't outsource all your core competencies. IDEO is in the business of making stuff, and we wouldn't be the same if our magic elves weren't on staff. IDEO's machine and model shops are our in-house alchemy labs, the place where proverbial napkin sketches take physical shape, where model makers, machinists, and skilled craftsmen interpret and stretch the boundaries of design and engineering. Some of our best work happens when an open-minded IDEO engineer or designer leaves enough room for the shop to play.

The Technologist

Every company needs someone who has read all the latest journals, can unlock Byzantine problems, or, when he doesn't know the answer himself, can point to seemingly obscure sources. In other words, a geek. We've got a resident technologist. His knowledge of complex materials, mechanisms, and seemingly arcane technologies provides a wealth of raw material for IDEO engineers and designers. He can't help but probe the deeper connections between things. Just by his presence, he can change the perspective on a project.

The Entrepreneur

Bright, innovative, but didn't quite find a niche within the firm. We've all encountered such people in large organizations, and if truth be told, it happens on occasion at IDEO. The solution for our entrepreneur was a spin-off that's since come back into the IDEO fold. It's a quirky, positive influence on our culture. Our entrepreneur constantly experiments with how to quickly brainstorm, prototype, and communicate innovations. Quite by accident, he's become one of our masters of process and a brainstormer extraordinaire.

The Cross-Dresser

We all know this kind of individual. Enthusiastic, self-educated, and self-motivated, cross-dressers have learned so much since they left school ten or twenty years ago that it might as well have been a past life. Their thirst for their chosen occupation makes them ideal team members. At IDEO the typical cross-dresser might have studied engineering but fallen in love with design. We've got so many cross-dressers that it's almost the norm. You know you're open to cross-dressing when team members' formal training doesn't matter, when energy, abilities, and personality are the main qualifications.

Teams may be ground zero for innovation at most companies, but they start with individuals. I'm sure you've got plenty of extraordinary people in your organization. Match them to projects, challenge them, and give them a chance to blossom, and I think you'll be surprised at the results. There's nothing like a hot team to get the job done.

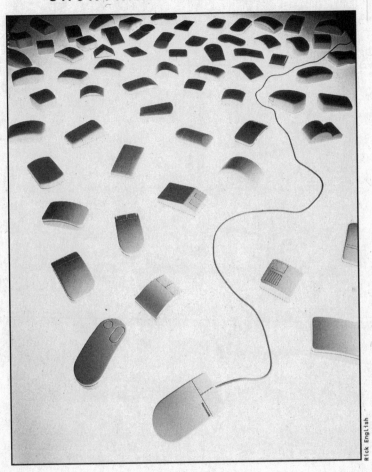

Rick English

The problem seemed insurmountable. We'd designed some new goggles for snowboarders, but weren't sure if our sleeker, face-hugging style would stay fog-free in the freezing conditions found on most ski slopes. We needed to test the goggles in the wild. But it was the middle of summer in sunny California. Our client at Smith Sports Optics didn't have the time or the budget to fly us down to New Zealand, where it's winter in July.

We couldn't wait five months for winter to come around, but there was another possibility. Two blocks south of IDEO's Palo Alto offices stands a classic family-run ice cream factory. The folks at Peninsula Creamery thought us a bit crazy when we floated our wild request for a nontraditional use of their industrial-sized freezers, but as long as we didn't eat all the ice cream, why should they care?

Industrial design studio head Paul Bradley and a handful of team members bundled up in parkas and long johns. The exercise bike and fan (to simulate wind) were found by tapping out an e-mail (the way all sorts of unusual requests get filled at IDEO). Since nearly everyone at IDEO is a bike enthusiast, it wasn't hard finding volunteers to pedal for an hour in an ice cream freezer, as observers snapped photos of their perspiring faces and, fortunately, not-too-foggy prototype goggles.

Prototyping is problem solving. It's a culture and a language. You can prototype just about anything—a new product or service, or a special promotion. What counts is moving the ball forward, achieving some part of your goal,

Not wasting time.

Prototyping is a state of mind. Think of the best moviemakers. We don't idolize them just because they strike us as glamorous players in a profession filled with beautiful people. They're also people who *get things done*. People who make movies prototype every day, just to get by. Consider the following story. Larry Shubert of IDEO worked for a few months on the set of the underwater thriller *The Abyss*. If you haven't seen it, all you need to know is that most of the movie happens beneath two thousand feet of water. That critical fact forced the *Abyss* team, led by *Titanic* director James Cameron, to seek out a gigantic body of water to film the action. A pool wouldn't do, but what about a half-finished nuclear power plant in South Carolina? The main containment vessel was a tremendous 250 feet in diameter and 50 feet deep. Cameron turned it into the world's largest freshwater tank—almost 7 million gallons. This being a film, there wasn't time to plan. There was also no time to build scaffolding, so parts of the interior set were created literally as the water poured in, with workers in rowboats furiously painting the walls as the water rose.

That's the right metaphor. No matter your business, no matter your experience, odds are the water's rising. You don't have time, and if you don't act soon, you—or your project—will be underwater. There probably isn't time to do things the regular way—or by the book.

When Shubert and a couple of other IDEO folks were told they were flying to South Carolina to work on *The Abyss*, they suddenly realized they'd be working underwater and needed to be scuba-certified. One problem. The scuba course was going to take weeks. "We said, 'There just isn't enough time,'" Shubert recalls. "'We'll go through the books on our own.'" After a little negotiation, they made a deal. Schubert and the others read the books, did the dives, and passed the test—in a couple of days instead of a few weeks.

KID STUFF

Many of us first learned about prototyping as kids, doing class projects or, to help pass those long summer afternoons, making up games. My brother David, for example, has been building things and then trying to make them better for as long as I can remember. We had a snowy Ohio winter the year I was six years old, and David started a series of increasingly complex snow construction projects in the backyard. He started with the basics—three-tiered snowmen—but soon progressed to whole forts by lining up snowmen shoulder-to-shoulder to form four walls.

Looking for the next revision of his prototype fort, David briefly considered a two-story model, which he thankfully abandoned when he hit upon the idea of using a cardboard box to make snow "bricks." We were industriously building snow fort 2.0 with our adobe-style construction techniques when David hit upon an idea for revision 2.1: adding water to each brick so that it would freeze to a solid (and incredibly heavy) block of ice, which David hoped might help the fort last until Memorial Day.

You're probably coming to the conclusion that my brother was incredibly driven from an early age, which may be true but is not the point. With our little team, we built the mother of all snow forts that year, and we did it by building a bunch of prototypes and then figuring out how to make each one better than the one before. David's gift was his faith in the process. He knew that if he kept trying out new techniques he'd make dramatic improvements.

I think a lot of us understood this intuitively as children and lost it gradually as we matured. David's influence has made this childlike curiosity and enthusiasm second nature at IDEO. A playful, iterative approach to problems is one of the foundations of our culture of prototyping. It can be part of your workplace too.

BUILD TO LEARN

Veteran IDEO studio head Dennis Boyle approaches client meetings like mini-shows. He often says, "Never go to a meeting without a prototype," which we now jokingly refer to as Boyle's Law. In fact, Dennis often brings *several* prototypes, as well as interesting materials to spark the muse. It's part of the iterative process. "We kind of shock people by how many things we'll build sometimes," he says. When a project's just getting off the ground, client meetings may be weekly. Dennis wants to make his mistakes—and discoveries—as soon as possible. As he puts it, he looks for any opportunity to make models, sitting down with his team and saying, "OK, what are we going to build for next week and the week after and a couple of weeks out?"

There's always something that wasn't clear at the last meeting that you can clarify by offering distinct choices, such as showing how you might implement a hinge or mount a display or control a specific mechanism. When the project is especially complex, prototyping is a way of making progress when the challenges seem insurmountable. "In engineering, if you have more variables than equations, you normally have an unsolvable situation," explains Dennis. "But on the other hand, if you just take some good guesses and fill in the blanks on some of them, you'll get some answers."

Focused prototyping helps resolve little critical problems one by one. IDEO studio head Sean Corcorran's mantra is "build to learn." When he was working on a chair for Vecta, for example, his team reached a point where a height adjustment lever that tilted with the chair became critical. They didn't build the whole chair. They didn't even build the whole tilt mechanism. They just built the little lever and its interface with the release mechanism. It only took a couple of hours. The finished prototype fit in Sean's hand and it quickly demonstrated that the principle would work.

HOW AMAZON DID IT

After reading about prototyping movies and goggles and snow forts, it's understandable if the process still seems undefined. "All well and good," you may say, "but how does this help me if innovation isn't a daily ritual? And how do I prototype a new service or business?"

Quick prototyping is about acting before you've got the answers, about taking chances, stumbling a little, but then making it right. Consider Jeff Bezos's story of the birth of Amazon.com. The business rookie launched one of the Internet's first multibillion-dollar behemoths literally on the run.

Bezos's saga puts wings on the Silicon Valley cliché of starting your company in a garage. His e-commerce juggernaut began with one startling statistic. In the spring of 1994 he happened upon a prediction that annual Web growth would ramp up at an astounding 2,300 percent. Inspired by the market potential of such explosive growth, he madly scratched out a list of things he could sell on-line—everything from music to clothing—before settling on books.

Within weeks he quit his cushy Wall Street job, called a moving van, and "packed and had our things taken before the van knew where they were going." Incredibly, Bezos hadn't yet figured out where to cast his e-commerce seeds. His short list included Boulder, Portland, and Lake Tahoe. Unable to make up his mind, he instructed the moving van to simply head west. The next day, Bezos phoned and told the driver he'd decided on Seattle, a city with plenty of high-tech workers.

Bezos was doing exactly what innovators do every day: breaking a problem down into its parts, making on-the-fly decisions in parallel. The ordinary thing to do would have been to stay put until he had decided upon a city. How could a moving van begin its journey without knowing a state, let alone an address?

But Bezos gained a day of Internet time by launching his ship before he'd charted his New World. As his moving van rolled westward from New York on the interstate, Bezos flew to Texas and picked up a beat-up family car. While his wife, MacKenzie, drove, he sat in the passen-

ger seat, pecking out a business plan on his laptop and punching out calls on his cellular. Think of it: nothing on paper, no place to land his imagined company, yet he was already hurtling toward his destiny. He took a detour through Northern California to interview potential vice presidents of development, and he retained a Seattle attorney by phone to incorporate his on-line venture with the unlikely name of "Cadabra." He still had no idea where he was going to live or how he would fund his scheme, but he had no time to waste because "when it's growing 2,300 percent a year, weeks are important."

Bezos's story would make a great movie about how rapid prototyping can give you a business edge. The dynamo entrepreneur was sorting out the pieces of the puzzle as he sailed west, trying to figure out what his Website would do, where he'd be, whom he'd hire—and what he'd call the darn thing.

Bezos could make an early blunder—like the clunky Cadabra—because he was carving himself out so much extra time. Once he'd rented suburban Seattle digs to house his venture in the proverbial garage, he applied the same "time is precious" logic to Amazon.com. Get it up, get it out, was the order of the day. Function preceded style and editorial content. Low on graphics and animation, Amazon.com loaded fast and excelled at the basics—making it easy to find and buy books.

By the time the dust cleared, the bricks-and-mortar booksellers were playing catch-up. The next time you kick off a project, try sparking your engines with a little New World Amazon-style energy. Think about tackling problems when you don't have the answers. Once you get in gear, you'll be surprised at how easily some of the solutions appear.

MAKE YOUR LUCK

Prototyping doesn't just solve straightforward problems. Call it serendipity or even luck, but once you start drawing or making things, you

open up new possibilities of discovery. It's the same method that's helped scientists unlock some of the greatest secrets of nature.

"I decided that no harm could come from spending a few days building backbone models," said 1962 Nobel Prize winner James Watson, referring to the metal prototypes he and fellow geneticist Francis Crick used to model and test their hypotheses. "Perhaps a week of solid fiddling with the molecular models would be necessary to make us absolutely sure we had the right answer."

Although their academic colleagues were skeptical of the approach, Watson and Crick's freethinking style and openness to two-dimensional and three-dimensional prototyping helped guide them toward the momentous discovery of the structure of DNA. Even Watson sometimes wondered if he was floundering. "My doodling of the bases on paper at first got nowhere. Not until the middle of the next week, however, did a nontrivial idea emerge. It came while I was drawing the fused rings of adenine on paper."

Doodling, drawing, modeling. Sketch ideas and make things, and you're likely to encourage accidental discoveries. At the most fundamental level, what we're talking about is play, about exploring borders.

Some of the bravest prototypers in the world are the people who build experimental aircraft—and the test pilots who fly them. The Bell X-1 aircraft flown by Chuck Yeager to break the sound barrier in 1947 was sometimes referred to in news stories as "a bullet with wings." What's lesser known, however, is that the first prototype actually *was* a bullet. When you're creating something new to the world, you can't look over your shoulder to see what your competitors are doing; you have to find another source of inspiration. During development of the historic aircraft, designers at Bell examined a fifty-caliber bullet flying at supersonic speed. The bullet was aerodynamically stable even as it broke the sound barrier, so Bell's engineers created a thirty-one-foot-long fuselage in the shape of that bullet, adding some stubby wings and a powerful rocket engine. Thanks to the brave prototype testing by men like Yeager—there wasn't even an ejector seat in the

X-1—we gained knowledge about supersonic flight that led to the creation of production aircraft like the Concorde and modern jet fighters.

Of course, we don't pretend to be Chuck Yeager or Nobel Prize winners, and IDEO makes no claim to discovering anything remotely as important as the structure of DNA. But the process of prototyping can spark little innovations, the sort that can be the difference between a product's success or its failure.

When designing a new video game controller for Logitech, we sent drawings to IDEO's machine shop to have them knock out a prototype steering wheel to be made of rigid ABS plastic with a black rubber coating. But the shop was completely out of black rubber, which, with a deadline looming, seemed like a potential disaster. So they molded the steering wheel out of red rubber, the only kind they could lay their hands on in time. By chance, the client loved the brightly colored wheel, and the little accident spun the product in a new direction, inspiring the game maker to introduce a line of fire-engine red steering wheels.

Brendan Boyle, Dennis's brother, invents toys at IDEO's Skyline group and pitches them to the major toy companies. One of his more successful discoveries came while gluing together the prototype of a toy football. Brendan and his team had hatched the idea of creating a foam-rubber football that had built-in tee-wings that kept the ball propped up so you could automatically tee it up for placekicking. Charlie Brown would have loved the idea.

Brendan fashioned his first prototype the quick and dirty way. He simply took an existing football and set about gluing wings on it. But as he began applying the wings, he saw that it would be easier to follow the existing curved line of the football.

"Oh, this is kind of cool," Brendan remembers thinking, seeing that if his wings traced the curves, it would look like a propeller. He quickly "punted" on the idea of a self-teeing football. Once he started tossing balls around the office, it became clear that the curved wings had a more noble purpose—straightening wobbly throws into perfect spirals. And so the Aerobie football was born, one of Skyline's most successful toys.

Daniels & Daniels

Failed prototypes opened the way to success
for the Aerobie football.

Get in the habit of making stuff or airing trial balloon ideas. Odds are you'll bounce your way to the insight that may inspire your next breakthrough.

PROTOTYPES BEAT PICTURES

Prototypes aren't only about making giant discoveries like DNA or little ones like a hot-selling toy. There's also something wonderfully tangible about a prototype.

It's easy to reject a dry report or a flat drawing. But models often surprise, making it easier to change your mind and accept new ideas. Or make hard choices, such as forgoing costly and complex features. Years of experience have taught us that prototyping is also part performance, that if the act isn't well orchestrated and substantial, the audience gets antsy. And executives, understandably, already have enough on their minds.

Give your management team a *report,* and it's likely they won't be able to make a crisp decision. But a *prototype* is almost like a spokesperson for a particular point of view, crystallizing the group's feedback and keeping things moving.

We believe in that great old saying, a picture is worth a thousand words. Only at IDEO, we've found that a good prototype is worth a thousand pictures. Somehow you up the data rate. Give people two or three very concrete choices. Create a situation where the audience can hash out ideas, and you can run with them or show why they might not fly. Bring lots of hardware in that shows the size and shape of things, how parts fit and work together, and people will likely say, "Oh, I never thought of that."

What kind of prototype works best in different situations is itself a moving target. Good prototypes don't just communicate—they persuade. If you're developing a consumer product, you might prototype out of foam or plastic or wood. But there are many other ways to prototype. Sometimes we quickly film what we like to think of as movie trailers, to show the highlights and essence of what a product, service, or business may become. If you're working on a project that has a service or human component, sometimes it helps to have team members—and even clients—express the project through archetypal characters in a little improvisational skit.

Living, moving prototypes can help shape your ideas.

BIT BY BIT

All too often, we've seen how employees prepare presentations for upper management in large corporations. Someone is given a few months to come up with a new product, service, or marketing campaign. More often than not, they don't solicit enough input or feedback early on—there are no incremental reviews. When the boss finally sees what

they've been working on all these months, it's either a happy surprise or a total disaster. Try getting in the mind-set of a quarterback facing a two-minute warning from the moment you start on a new project. Instead of the long ball, try to get a few yards, passing to the sidelines to stop the clock. Just keep the momentum going.

We pitch presentations in stages, show the rough sketch, the cheap foam model, and use them to right the course before it's too late. As Dennis says, if you only have till next Thursday, you actually have to decide on some part of the product. You've got to cut and run and tell the shop to work up a model.

Or then again, you might buy your prototype from a dime store. Twenty years ago IDEO veteran Jim Yurchenco bought part of a key prototype for the first Apple mouse off the discount shelves of the Walgreens store across the street. A butter dish bought for a couple of bucks was just the right size to hold the mouse's rolling ball. That isn't to say that we didn't make dozens of subsequent prototypes. It just shows that what counts is expressing the idea quickly—and cheaply.

This isn't a concept that only works for start-ups. A few years back, Wells Fargo bank was struggling with a feeling we know all too well. If someone shadows you while you're tapping away at an ATM, you probably get uncomfortable or even fearful, especially if you're withdrawing a couple hundred dollars and the streets are deserted. So when IDEO took on the bank's project, we observed how people behave at ATMs and began brainstorming all kinds of ideas, including expensive add-ons like video cameras and periscopes. But it turned out that sticking a $1.80 fish-eye mirror (the kind that truckers stick on their big side mirrors) above the ATM worked just fine. Sure, we could have pushed them toward more high-tech, expensive solutions. But prototyping can also remind you that sometimes the most obvious, simplest solution is the best.

SHOOT THE BAD IDEAS FIRST

Prototyping is a dance. Sometimes the music doesn't move you or your steps fail. But that's no reason to stop. Just as writer's block happens when writers stop writing, so, too, does innovation grind to a halt when prototypes stop being built. When the muse fails you, don't mope at your desk. Make something.

"I just prototype a bad idea and maybe shoot the hell out of it," says one of our Silicon Valley–based engineers. He'll often know it's not the solution he wants, but if he prototypes it, he can shoot it down faster and then find out what doesn't work or perhaps discover something new. Prototypes can be a source of creation and insurance. When all else fails, prototype till you're silly.

HOW APPLE DID IT

It's easy to rhapsodize on the value of modeling. But sometimes only a real-life story will show you how the process actually works. And few products better demonstrate the importance of quick prototyping than the Apple Duo Dock. The time was the early nineties. Apple wanted to introduce a portable that could easily dock in a home base with a standard monitor and keyboard. The portable had to be small and thin, and the company initially anticipated some sort of mechanical dock using levers they'd already seen.

It bears remembering that for decades—long before the debut of the iMac—Apple had been a leader in industrial design. In the early eighties, without Apple the profession almost wouldn't have existed in Silicon Valley. PC clones were manufactured, not designed, and we owe a debt of gratitude to this colorful company which always recognized that behind every transaction was a real person.

But with the Duo Dock, Apple had the usual pressures. It's natural to be skeptical of new ideas, especially when they're likely to cost more

and increase risks. It seemed like the laptop would dock mechanically, but we tried not to let that constrain our brainstorm. We were search-ing for a metaphor. At the time, Dennis Boyle had a three-year-old who'd already figured out how to pop a video into a VCR. If the TV was already on, it would just start up and play his cartoon. Didn't that kid-friendly experience fit perfectly into the tradition of the Mac graphical interface, the wonderful, playful simplicity that was Apple's trademark?

What if we could do something that direct—and powerful? What if you could just pop the Duo into the dock like a video—and it would sweep in effortlessly and play like a movie? Though the idea sounded promising, Apple at first was reluctant to take any chances. The com-pany preferred a traditional mechanical eject system. The VCR ap-proach sounded expensive, like something that might delay the launch date.

But the VCR metaphor stuck with the IDEO team. Dennis and company brainstormed again. They quickly hatched (and later patented) their key innovation, which involved combining a toy motor with a clever gear mechanism. The other option, using traditional sci-entific motors, was pricey—anywhere from ten to thirty dollars. If they'd used the sort of precision motor people expected them to use, cost alone would have killed the idea. But toy motors are dirt cheap—about $1.25—and surprisingly reliable.

Dennis gambled. He temporarily suspended work on developing a mechanical lever and told the team to make it with the toy motor. A couple of weeks later he brought in a working model to Apple. They were so excited they took it around to everybody and got instant buy-off. Sure, it cost a little more and would take a few weeks longer. But Apple recognized that the simplicity and smoothness of the VCR-like approach would give it a competitive advantage. And more data pro-tection. One risk with mechanical docks is that users try to eject their computer before files or the operating system software are ready. Since the motorized approach was itself software-controlled, the Duo would only eject when ready.

Still, there were problems to solve. When early models spat the Duo

out so forcefully that the expensive laptop looked precariously close to crashing to the ground, we added an oil-filled damper to soften the movement. More than a dozen different prototypes were made. Early models were fashioned of particleboard and ABS plastic, testing levers, and soon the toy motor mechanism. They were often cobbled together in an afternoon.

The working prototype itself became a celebrity, like a popular exhibit model at a science fair. These were the days when Apple CEO John Sculley would stroll through Apple's labs to get a firsthand look at the latest projects. This particular day, he was supposed to review five different projects in about half an hour, but once he spied the Duo Dock prototype, he didn't budge. The other projects had static charts or simulated images on the screen. The Duo Dock was entertaining. You could open it up and see its clever mechanisms working away. Sculley wanted to know all about it, watching, listening, and talking about it.

Nearly every prototype sparked improvements during development of Apple's Duo Dock.

Apple ordered a couple dozen copies of the next prototype so it could demonstrate the concept to customers and partners before the real ones rolled out of the factory. In the few weeks before the new prototypes arrived, the original became so popular that it required its own conference room. You had to schedule a meeting if you wanted to see the prototype. The Duo Dock earned great reviews, but we're equally proud of how cheaply Apple was able to manufacture the innovation. Fear of added expenses nip countless innovations in the bud. But the toy-motor-powered mechanism embodying the VCR approach required less than seven dollars in parts—roughly comparable to the cost of some purely mechanical solutions.

To me, the Duo Dock is a classic case study of why an iterative approach works. Prototype with energy and enthusiasm, and you've got a good chance of hitting upon the very feature or product that resonates with your customer.

FORGET CONGRESS

Okay, so maybe you don't make products and are having a hard time seeing the value of prototyping for your business or service. You don't believe the shoe fits for every industry. What about a national park? If the federal government can prototype on the fly, isn't that a good sign that even the most daunting bureaucratic obstacles are no excuse, that with the right energy and ideas anyone can innovate?

In 1997 the Golden Gate National Recreational Area was struggling like a geriatric multinational corporation, unsure of its identity and future. A patchwork of roughly two dozen sites strewn along seventy miles of rugged Northern California coastline, the parks lacked funding and volunteers, and—despite crown jewels like Muir Woods and Alcatraz—many of its riches were barely known to the public.

Board member Mickey Drexler, chairman of the Gap, knew they

had a problem. Nobody, not even fellow board member and advertising guru Rich Silverstein, knew exactly what the GGNRA was—or what it was selling. "We can't even define it ourselves," Drexler declared during a key powwow. "We can't even put it on a T-shirt."

The T-shirt idea set Silverstein (famed for his celebrity-studded "Got Milk?" campaign) to thinking. He brainstormed with his partner Jeff Goodby. They quickly recognized that the "Recreation Area's" run-on name stunk. Why not just change it? Well, for one thing, that kind of change requires an act of Congress.

Unless you break the rules. So they hatched the "Golden Gate National *Parks*," earned the ire of federal bureaucrats, and forged ahead. Following the Gap chairman's logic, they hired the gifted illustrator Michael Schwab, who created the marvelously simple "brand look" for the first of many icons—a sweeping image of the Golden Gate Bridge and the parklands on either side (never mind that the bridge itself wasn't technically part of the park). The icons were so hot that when they premiered in bus shelters people actually stole them. A barrage of phone calls from consumers made clear that plenty of others were dying to buy them. And so the name—and numerous classic images—burst forth and multiplied on T-shirts, sweatshirts, coffee mugs, and calendars. In just two years, the park raised $26 million in charitable contributions, launched a couple of profitable retail stores (you've gotta like a federal agency that invents profit centers), and recruited fifteen thousand new volunteers, helping to put the shine back on a national treasure.

The entrepreneurial, rule-bucking Golden Gate National Parks is now the poster child of the national park service, and the same bureaucrats who bemoaned its shoot-from-the-hip style now come to study the masters, treating the park like a living Harvard Business School case study. The best irony of all is that if they'd waited for Congress, they wouldn't have gotten past Go. The feds still consider the original, cumbersome "Golden Gate National Recreation Area" the official and legal name.

They can keep it.

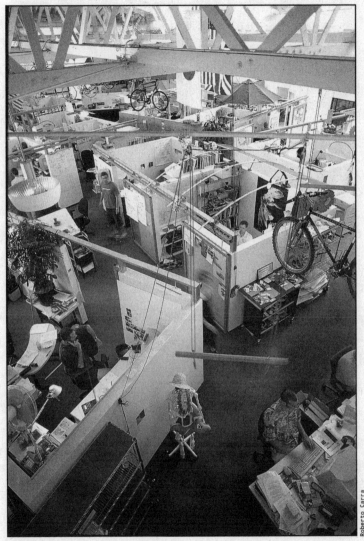

Roberto Carra

Prototyping, brainstorming, and observations. These are the fundamentals, the reading, writing, and arithmetic of innovation. Great teams provide the charge that makes these basic skills flow throughout a company.

But just as brainstorming doesn't exist apart from teams (no, we don't count solo effort like thinking in the shower as brainstorming), the process of innovation doesn't happen in a vacuum. Innovation needs teams. And teams need places to thrive and grow.

Innovation flourishes in greenhouses. What do I mean by a greenhouse? A place where the elements are just right to foster the growth of good ideas. Where there's heat, light, moisture, and plenty of nurturing. The greenhouse we're talking about, of course, is the workplace, the way spaces take shape in offices and teams work together.

Perhaps you consider office space secondary to the nitty-gritty process of innovation. Well, at IDEO we consider it one of our premier assets. A few years ago I was quoted in *Fortune* magazine in a cover story on "The End of the Job" as saying, "Hire the right people, and everything else will take care of itself." Life is not quite that simple, of course, but hiring talented people is the top job of every business unit manager in the world. And top people tell us they join IDEO not only because we lead our industry but also because of our physical and cultural environment. Our office spaces help make innovation happen.

If that still doesn't sway you, how about a little bottom-line argument? At most companies, office space is the second biggest expense category after salary and benefits. Every competitive company knows it has to pay well for top talent. Why shouldn't you pay the same attention to how you create the spaces in which this talent must perform? Athletes need proper facilities. Why not workers?

Creating great office space may be one of the hardest parts of the innovation puzzle. It doesn't happen simply by hiring forward-thinking architects or leasing cool buildings. Indeed, some of IDEO's best

spaces have evolved by eschewing any sort of overriding master plan. We take the opposite of the Dilbert approach. It's pretty simple. When it comes to office space, the fewer rules the better.

That said, we think we're pretty good at creating office space that works—and looks great. Along the way, we've developed some basic tenets that will help you in making your workplace better fitted to innovation. Start by building neighborhoods and thinking in terms of projects. Make building blocks to create playful, flexible foundations for your workers. Throughout the process, remember the spirit of prototyping. Space should evolve with teams and projects in much the same way as do your plans for an innovative new product. Encourage hot teams to find or create a team icon.

Finally, don't forget that your spaces should tell stories—about your workers and your company.

BUILDING NEIGHBORHOODS

Our offices at 700 High Street in Palo Alto have perhaps been photographed by more magazines than any other IDEO space. The flights of fancy are great, but the most important feature for companies trying to create spaces that support innovation are closer to the ground: the neighborhoods and personal touches. High above float the colorful umbrellas and national flags; a full-sized jib sails just a wave away from a snowboard with boots but no boarder; and in the corner glares a defiant pirate flag. At ground level, visitors are often struck by the plastic-beaded curtains some people use as doors or the Christmas tree lights that shine year-round. The main pathway leads past our murals and project displays, a wall that illustrates a medical project we worked on to more efficiently grow human skin for burn victims, and the GM Hughes charging station still used today for General Motors' EV1 electric vehicles.

We believe in the importance of neighborhoods and community in fostering innovation. Try creating spaces that draw workers in and encourage interaction. Our spaces, for instance, have been laid out so that a central asymmetrical table functions like a park for three or four team members. Prototypes, sketches, or blueprints crowd these tables, and if workers aren't hunkered down, their Palm Vs or cell phones usually signal they're nearby. Everybody who lives by that "park" can see at a glance whether their neighbors are at home, but they can also slide translucent Lexan barn doors closed if they need to buckle down and work privately on something. It's a wonderful mix of community and privacy that seems to offer much of the seclusion of traditional offices without the separation.

Our Palo Alto offices are great examples of how neighborhoods and communities can spill into the streets. We started in a small office in Palo Alto, just a few blocks down the street from Stanford University. Now we've got seven modest offices housing about 160 workers, all within a few blocks of one another. (IDEO also has offices in Europe, Asia, and sprinkled across North America.) There's a reason why we don't cram all our Palo Alto staff into the same building. Smaller offices make it easy for our studio teams to put their own stamp on their space. Each building's character and personality reflect its workers and their particular blend of projects.

Spaced along pleasant, treelined High Street, just off Palo Alto's bustling University Avenue, the offices—and our workers—enjoy the local buzz. We bump into one another when going out for a sandwich or a cappuccino, small clusters forming outside an office. IDEO is a microcosm of a college campus, with ample opportunities for serendipitous encounters, as well as secluded, focused work. Since most of our spaces lack internal halls, High Street has been called the "hallways of IDEO," the places where spontaneous interactions take place.

We aren't alone in looking to turn a walk around the campus into an innovation opportunity. Swedish telephony powerhouse Ericsson has made this idea of communication-on-the-go central to its new London

headquarters. An abstract, curvaceous staircase links floors, giving plenty of room for the company's handset-holding workers to stop and meet. Indeed, the company says the building's principal idea is to "increase the natural foundation for meetings between people."

But while grand, architectural gestures like these are important, it's the personal nature of neighborhoods that drives innovation. At IDEO people celebrate their space with prototypes they've worked on, found art, and clever mechanical toys. One of my favorite corners looks like something out of a bike shop. There's a hat rack full of cycling helmet prototypes, a plywood pendulum with a heavy metal ball to test them, a phenomenally light carbon racing wheel, and a clever folding bike lock of hardened steel.

At nearly every corner, mobile steel Metro shelves are brimming with foam prototypes or parts. If someone's got a hobby or interest, you'll find it in his or her space, everything from cell phone collections to the mishmash of horribly designed toys and gizmos. As for the bizarre, IDEOer Roby Stancel favors a Styrofoam dress shop model's head to store and display his electrical components (sort of a cyberpunk hairdo). There are also plenty of surprises: quirky wall cutouts and window slits that afford unexpected views and provide amusing displays for knickknacks.

Look around your office. Do workers have places for impromptu meetings? Is there a comfortable blend of openness and privacy? And equally important, are workers encouraged to celebrate their work and hobbies in their space? If you can't tell when you're moving from one neighborhood to another, you probably don't have neighborhoods.

THINK PROJECT, THINK PERSONAL

Visitors are often struck by the anything-goes, kindergarten look of our offices. But our innovation edge isn't the color, wild materials, and

toys. Plenty of hip, young Internet companies aren't exactly gray. The reason our offices seem to work is that nearly everyone has a say. We're firm believers in the hands-on approach. For example, our studio heads often design the overall pattern of their space and then let team members mold their own nooks and crannies as they see fit.

While IDEOers are encouraged to sketch out their identity like finger painters, we prefer watercolors to oil paints. Everything from office partitions to files and desks are easily movable. We've designed and built little carts to neatly cluster the web of power and telecommunications wires. When a team member moves, it's as simple as disconnecting and pushing the cart to their new location. We've got a shrink-wrapping technique so that you don't even have to take things off the shelves before you roll them away. Space is the team and the work. If a member wants to jump aboard another project, he or she needs to be able to quickly take off and land in another building or office.

How do you keep space evolving and elastic? One way is not to take your space or yourself too seriously. Just as IDEOers really don't earn a traditional window office by achievement or rank, our culture frowns on the idea that anyone's status or space is sacrosanct. How wild can it get? One studio head returned from vacation to find his office entombed in a plaster wall. Still another returned from Hawaii to find his desk and files sitting on a custom-built redwood deck, complete with beer cooler, patio table, and plastic pink flamingo. And another returned to find his transformed into an "executive washroom," with a linoleum floor, shower curtain, and a toilet where his office chair used to be.

This may sound a bit extreme, but companies that depend on the creativity of their staff give them free reign when it comes to space. At Pixar, one worker's station resembles a tiki hut, complete with bamboo door and roof; another evokes a child's bedroom, with oversized stuffed animals and a bunk bed. Why should businesses encourage such playfulness? Well, Pixar's freewheeling, independent approach to space has helped the company make some pretty darned creative block-

busters. Their first three feature films, *Toy Story, A Bug's Life,* and *Toy Story 2,* have collectively topped a billion dollars at the box office, and that doesn't count all those toys and licensing deals.

But I'd be the first to agree that most companies don't see the connection between creative space and innovation. We were recently on a British television show that cut back and forth between David's office and the office of a traditional Fortune 500 CEO. The contrast was dramatic and spoke volumes about the cultural differences between the organizations as embodied in their physical spaces: David's office with a thousand eclectic items, the flotsam and jetsam of twenty years in hands-on product innovation, versus the stately CEO suite with a large couch and blank walls: no computer, no papers, not even neatly filed memos, just one of the company's antique consumer products carefully positioned on an otherwise empty credenza in the background. At IDEO we're never tidy, but all the mess is not madness. When David wants to make a point to a visitor, he's like a well-outfitted puppeteer—all the props are within arms' reach.

BUILDING BLOCKS

Building blocks are fun. They become part of your company's space toolbox. They actually help you innovate.

The cheapest, simplest materials can inspire workers and become a building block that takes shape throughout your offices. Take these milk-crate-sized foam cubes you'll find throughout our offices. By chance, Douglas Dayton, who heads up our Boston office, did some acoustics consulting in a former life. To dampen the clatter of noisy heaters and air conditioners in IDEO's brand-new Boston office, he arranged to have a local firm make thirty gray soft-foam cubes about fifteen inches high. He wasn't thinking only of function. The building's icon is a geometric cube sculpture out front that represents IDEO's

Dean Powell

Simple foam cubes can contribute to the playfulness
and versatility of a work space.

four-square logo. With a little glue and piano wire, Douglas fashioned
his thirty cushiony cubes into an acoustic "cloud" over his desk that
mimicked the sculpture.

The cubes were contagious. He bought three hundred more, origi-
nally intended for creating more "clouds," but they quickly disappeared
throughout the office, becoming makeshift chairs, tables, and room di-
viders. Still another three hundred were instantly absorbed. Soon there
were more than eight hundred—in an office with only forty people!

Cubes make great impromptu conference room partitions,
amphitheater-like seating (just back them up to walls), and wonderful
entertainment for visiting kids. And in a clear example of how creative
space can feed your creativity engine, the cubes have even helped us
innovate. To prototype an elevator for Otis, the world's largest elevator
manufacturer, the Boston office stacked cubes to simulate the interior

of an elevator so you could actually walk in. It was perfect for testing how fast it takes to load and unload, and it was great fun.

INSPIRATION FROM ADVERSITY

Space problems can actually inspire innovation. The high-ceilinged, sunlit offices at 700 High Street, for example, are both a joy and a challenge. The light makes for lots of glare on computer screens. In the old days, most companies would have ignored the problem or imposed a heavy-handed solution. Instead, we held a contest to see who could come up with the best idea for diffusing the light. One designer took a trip to a home-improvement store and cleverly created white arches from PVC tubing, held in place with galvanized plumbing fixtures and covered with window screens. His winning entry earned him a great Italian desk lamp, but the "people's choice" solution was a simpler, far less expensive, arguably more attractive solution—oversize green market umbrellas, which sprouted like psychedelic mushrooms from the ceiling. Space can be both a resource and a place for fun.

Encourage people to solve their own space problems, and you'll likely spur innovation where you didn't know it existed. Years ago Jim Feuhrer, one of our machine shop veterans, found there wasn't room for his bike, so he hung it from some overhead hooks. Faced with higher ceilings in another building, another IDEOer, one with extensive sailing expertise, devised a clever system of hooks and pulleys to hoist up his wheels. No one complained. No one was reprimanded. Today, so many bikes hang from our high ceilings that you sometimes get the feeling you're in a bicycle factory.

Is there a space contest you might hold? Do the limits of your office space offer opportunities to stretch the innovative muscles of your team? Think of the interest people take in their apartments and

homes. Give people ownership in their workplace, and you'll be sur-
prised at the positive results.

PROTOTYPE YOUR SPACE

At its best, space inspires and amuses. Our Boston office's foam cubes
became both process and play. Employees felt elated after crowding
onto and off the simulated foam elevator. The merging of fun and work
was invigorating.

But our building and testing of a foam elevator wasn't anywhere
near our wildest experiment in space and process. Though we're
trained to build prototypes we can touch and feel, even we recognize
that some things are too big to prototype. Well, in Boston we decided
to ignore that logic. We managed to build a nine-foot-high, forty-two-
foot-long prototype of an entire Amtrak train car—and get it in the of-
fice.

Our Amtrak design team started working in the car. It was where the
project lived, where the team members were trying to soak up the ex-
perience they were trying to create. You couldn't walk by it without
finding three or four people pacing around, sketching, sitting down,
asking questions. The space invited experimentation. In a couple of
hours, we could install new spotlights or check ambient lighting. Situ-
ated next to our café, conference room, and machine shop, the train
car became *the* place to hang out at IDEO. People held meetings in
the car or tapped away on laptops. The train became a stage that made
possible plays we'd never imagined. We split it into a mock-up of a
"bistro" café car and a passenger car.

Amtrak brought a dozen conductors, service attendants, and man-
agers to walk around—and at times upwards of twenty people were
tromping around the structure, our own jungle gym. Since nearly

everything was built in foam core, we could often model a suggestion that same afternoon. Amtrak could actually get buy-off on the design by walking key internal audiences through the car. The train was prototyping, space, process, furniture, fun, and one heck of a conversation piece. Today you can experience the real thing in Amtrak's Acela express service from Boston to Washington via New York.

Space projects offer your workers great opportunities to expand their notion of office, work, and play. Think hard. Is there a way to integrate prototypes or the prototyping process into your company's work environment?

CREATE A TEAM ICON

The Amtrak train car served for several months as a kind of icon of innovation for the Boston office. But it's not always possible to turn part of a project into inspirational space. The best teams and studios make a practice of seeking out icons that inspire the workplace. The more adventure and camaraderie you can attach to the icon, the better.

One of the greatest IDEO icons began as a treasure hunt. A young IDEO group called the Spunks came up with the whimsical team-building idea of searching out a great centerpiece symbol for their space, akin to the pirate's flag Steve Jobs flew over Bandley One to mark the Macintosh development team's territory. The Spunks rented a bus and drove down en masse to Thompson's Aviation Library in El Monte, California, a graveyard for airplanes in the sands of the desert near Los Angeles. Most of Thompson's business was renting crashed airplanes for Hollywood films, and the IDEO contingent had a wild time stumbling around the great aviation relics. They were looking for something a little less than twenty feet long and found just what they'd dreamed of: a wing from a vintage DC-3, the same type of aircraft that flew Ronald Colman to Shangri-La in the 1937 movie *Lost Horizon*.

Back at IDEO, after a lot of weekends spent buffing and polishing, the wing made for a pretty wild room divider, but its most interesting use came later. When Peter Skillman, the ringleader of the treasure hunt, joined Dennis Boyle's studio at 660 High Street, he knew the wing needed to fly over a central spot. He chose the popular conference table adjoining the studio's tiny kitchen, and had our model shop build a full-scale foam-core and wood replica. They had to prototype the wing, you see, hang a lighter replica from the ceiling by cables to test the forces the real wing would likely exert. Space became an exercise in work and play, and the result—a dramatic soaring wing—is now a wonderfully expansive metaphor to inspire the studio.

The team icon doesn't have to be big. It can be a little thing. Or more precisely, lots of little things. OXO International, famous for its oversized Good Grips kitchen gadgets, understandably has an interest in how people of all sizes and shapes hold its products. One day in New York, Alex Lee, the president, noticed several discarded gloves on the sidewalk. Intrigued by these small pieces of "found art," he asked his staff to start collecting gloves. Today the company has an amazing, eclectic collection. It's not just a good conversation starter for visitors. The gloves represent the range of hands OXO's products must fit, and provide a living, breathing symbol of the firm's mission.

Are you launching a new team? Maybe there's an artifact that might inspire it. Or something you might actually cobble together or adapt for your office. Turn creating your space into a team-building exercise, and you'll be surprised at how much fun and pride people take in their workplace.

WATCH YOUR BODY LANGUAGE

As your office space starts to take shape, it's a good time to take notice of your "body language." Franklin Becker and Fritz Steele in their book

Workplace by Design talk about the physical work environment represented by a company's office space as a kind of body language for the organization. As we all know from personal experience, however, verbal and body language don't always agree. What is the body language of your organization saying to employees, clients, and visitors? Does your company send signals of hierarchy, or is it virtually egalitarian? Is it a place of radiating possibilities or faded glory? Ericsson, for instance, believes there is a connection between its cool new London offices and the company's "progressive" attitude.

Is your company sending the right signals, or is its office space contradicting its corporate policies? Becker and Steele claim that an R&D center laid out in rigid ways and using delicate, easily damaged surfaces is like a lover whispering "I love you" in a bored voice while doing a crossword puzzle. Watch your company's body language. It's speaking to you even if you haven't been listening.

As a management consultant, I've visited my share of offices that send out confused signals. Not too long ago the window office, walnut desk, and leather chair were nearly as important as the salary in reflecting executive status. In this carrot-and-stick model, only the champions earned leather and wood. But I know firsthand that lots of folks rebel against that work model. I still remember the day I interviewed at the consulting division of one of the big accounting firms. Great people. Reputable firm. Reasonable salary. It all looked pretty good until I asked to see what my office space would be like as a new associate.

First, my simple question about space flushed out the truth about lifestyle. Suddenly, I heard that you travel so much you don't really need a desk, which was a different story than I had heard earlier. Second, I got a clear signal about how they valued new associates. Not only did I not get a desk, I didn't get a filing cabinet or even a drawer. Where would I store my files and things? I'd share a gray sheet-metal utility shelf with another associate, my portion marked only by my name scribbled on a piece of masking tape. Maybe it wasn't quite rational that I turned down the job offer partly because of a gray utility

shelf, but the space had communicated a lot about the firm's culture, and I wonder how much talent was fended off over the years because the "body language" of the office said, "New associates don't matter."

SIMPLE TEAM SPACE

When you're creating or updating space, remember an easy but often elusive goal: Keep it simple. It's a big deal to change the physical workplace, and companies can get as excited and carried away with new approaches to office space as they do with the latest management trends. That's why it helps to have a sense of the history of office space as well as the latest trends before putting your own unique stamp on the workplace.

This is one part of the innovation puzzle where you can actually see with your own eyes what works and what doesn't. We've come a long way. Remember the early-twentieth-century photos of endless rows of typists or clerks immortalized in movies like *Desk Set* and *The Apartment*? There was a time when control and supervision were the message, the factory was the model. Fresh ideas and teamwork weren't part of the program.

At the close of the 1950s, Union Carbide's new headquarters in New York City epitomized the rise of space as status. In the following years, the building became as hierarchical as a feudal fiefdom. Where you worked was what you were. Climb the corporate ladder and you'd be rewarded with more space, wood furniture, broader views, and closer proximity to senior management. Grades and colors of carpet and wallpaper separated the haves and have-nots. The smallest detail could betray you. If your ashtray wasn't made of the right stuff, maybe you weren't either. It's easy to say that those were the old days, but some new ideas in office space can be as trendy—and impractical—as fashion. Several years ago, for example, some companies rushed to join a new fad—the totally open office.

Declaring war on private offices, Chiat/Day, the talented advertising agency, brightened spaces with flashy tables and chairs and mobile workstations dubbed "hives." The *New York Times* praised it as "a remarkable work of art," but workers felt lost. Nobody had a home or even a neighborhood. Each day, workers were just expected to grab whatever space was vacant. People pined for privacy and often stayed home if they needed to get something done. Confided one worker to the *Times*, "Creative people need quiet places to think, and that was something they never planned for."

Not surprisingly, in the mid-1990s when the Omnicon Group bought out Jay Chiat, the company chucked the totally open office. New management added some private spaces for writers and art directors, and created more project rooms and a room full of tables and computers for teams to spread out ad campaigns. In other words, the firm listened to its own workers.

It's an invaluable lesson in the critical role space plays in a company's work process and identity. Creating a great work environment may be nearly as important as hiring the right people. Indeed, in my experience they're often inseparable. When people interview at IDEO, you can often see their surprise and delight that our office doesn't look like a workplace. There are as many ways to achieve this as there are to lay out a house. But in the numerous companies I've visited around the world, I've found that the best offices all seem to celebrate teamwork. And sometimes that's about fairness and simplicity more than anything else.

The top high-tech company founders nearly always grasp this essential principle. Hewlett-Packard, arguably the first Silicon Valley enterprise launched in a garage, had an egalitarian approach to space for decades—almost nobody had an office. Jeffrey Bezos, founder and CEO of Amazon.com, intuitively understood the importance of frugal and fair office space. Bezos also started in a garage, and by the summer of 1997 when Amazon went public and had a few hundred employees, the company still held true to its spartan ways. Rent was cheap. Across the street stood a pawnshop, a wig store, and the Green Tortoise Hos-

tel. Bezos believed in the power of example, of managing and marketing by metaphor. He bought the best computers and hired the best people and scrimped on nearly everything else. The only sign announcing Amazon.com's headquarters was a sheet of paper stuck in a plant in the lobby.

Bezos's cramped shoe-box office was no bigger than his assistant's, and his desk, like most in the start-up, was an unfinished door with four-by-fours for legs. That desk was more than just a way to save a few hundred dollars—it was a symbol for workers, Wall Street, and investors, worth millions in marketing and good karma. Amazon wasn't merely an upstart challenging traditional retailers. It was short-circuiting the corporate status quo, and for its quirky mix of grunge workers and liberal literati it mattered that there was no walnut or mahogany for the executives—that everybody's PC sat on the same cheap door.

Visit the best start-ups, and you'll see a self-awareness that extends to space. Back in 1998, when Yahoo's market cap was already in the billions, more than a few Yahoo walls carried pictures and information about employee-inspired community efforts for youth and the environment. Even then, with dozens of multimillionaires, Yahoo walked and talked like a company that hadn't lost sight of its roots. One of my favorite touches was its executive offices. On the ground floor stood Amdahl, a dinosaur from the mainframe computer era. You got the sense that Yahoo was growing out of the ashes of the past, that the executives liked the humbling sense of walking into someone else's lobby before they reached their own.

Small or fast-growing companies present unique challenges. Start-ups, for instance, have revolutionized American business, but overlooked in the stock options and venture capital is that if you survive the first round of financing, you've probably outgrown your first or second building. Companies dripping in financing often just lease a cavernous building, leaving half or more of it empty, figuring that in the proverbial nine months they'll be bursting at the seams. But I've seen too many companies—even IDEO—lose momentum when there was too much extra elbowroom. The real secret is to crowd it with people, but it helps

a little to crowd it with furniture. When Handspring first moved to its building in Sunnyvale, it had a literal and figurative emptiness, but when I visited a few days later, the previously cavernous space had been filled with furniture, transforming it into more of a buzzing start-up.

I've seen the same "density-yields-buzz" phenomenon in our offices around the country. When the body count per square foot was high, the offices seemed to pulse like hot nightclubs on Friday night. It's a delicate issue to manage. No company wants to cram people into the office like sardines. But too much space, we've found, is like too much time for a deadline—we're all happy to take it, but it doesn't necessarily help get the job done.

HIERARCHY IS THE ENEMY OF COOL SPACE

As you create or rework space, beware of a natural tendency to reward superiors with superior space. Overly hierarchical corporate rules about space can drive a stake between you and your plans to innovate. Once, I spent all morning consulting with a well-respected Fortune 1000 company that said it wanted to shed its old bureaucracy, inspire hot groups, and learn to collaborate. Then they took me upstairs to the sumptuous—and exclusionary—executive dining room for a steak, and didn't see the irony.

If you've worked for a large company, you've probably witnessed firsthand how company policies toward space and hierarchy can drive people to absurd, petty lengths. Companies that rigidly assign office space based on rank can be the most abusive. IDEO's David Haygood, director of business development, for example, told me of a former employer who sentenced workers to dismal cubicles if they were "grade 19" or lower, while those at grade 20 and above got a hard-walled office with a door. Constant reorganizations created greater tensions. Once word leaked of an impending reorganization, grade 20s flew like vul-

tures to the part of the building where they'd be moving. Eyes on the ceiling, they dashed in and out of offices, counting. What in the world were they doing? By counting the ceiling tiles, workers could quickly ascertain the precise square footage of each office and knew for certain which office to get dibs on first. It's a miracle fistfights didn't break out.

Wise to this depressing company ritual, Haygood had the next company he worked for rip out the false ceiling and spray-paint the overhead mechanicals black. Next, since Haygood was in charge of office allocation, he assigned himself the worst office in the building—right next to the restroom.

The usual bickering ensued. Within minutes a few malcontents came to Haygood's dismal office, bitching and moaning. Haygood smiled at them and told them that anyone who didn't like his or her digs could have his. That shut up the complainers fast.

Ridiculous, right? Isolated case? Maybe. But I had an experience early in my career that says Haygood's story was not unique. With my newly minted MBA, I was ready to conquer the world the week I started with a global management consulting firm. What was my first assignment? Marketing strategy for a major airline? Operations analysis? No, actually, I spent the better part of a day making very precise measurements of each partner's office, including deductions for any support pillars or obstructions that decreased the number of available square inches. And why? One of the partners felt he had been slighted in the recent office move, and he needed my data to help argue his case and right this injustice.

Hopefully, such battles are on the way out. Over the last few years, a number of forward-thinking corporate executives are leading the charge against such status symbols. A senior executive with AT&T in Ohio chucked his office door and wood-paneled walls entirely and installed clear glass; not surprisingly, his management team quickly followed the lead. Throughout the nation, many corporate executives have abandoned their plush private offices and ventured into open-plan leadership communities. And even more have broken the umbilical cord between window views and rank.

If you're not the CEO, sometimes it can be tough to change the status quo. One midwestern multinational rewards executives by issuing them a leather chair when they enter the inner sanctum. One day, a new executive declared that he didn't want leather and ordered an ordinary chair. That tiny act of independence was viewed as a dangerous rebellion. Some executives worried that if "one of their own" rejected the leather privilege, soon other privileges would be snatched away from the executives. And so, incredibly, the executives insisted that the maverick sit in a leather chair.

GIVE YOUR WORKERS A VIEW

You and your company don't have to be a part of the leather chair set. You can literally give your workers a view. All of them. In the early 1990s, Alcoa realized that the old equation of space equals seniority thickened its corporate lifeblood. A revolution was started at the top, with Chairman Paul O'Neill ordering his executives into open-plan offices. The experiment worked—O'Neill liked how it encouraged his staff to commune in the kitchen and common areas—and so he charged ahead with his plan to hire an architecture firm to design a new headquarters near Pittsburgh's Three Rivers Stadium.

Today Alcoa's remnants of the 1950s—status-separating private offices and executive parking spaces—are gone. Hallways were chucked because they divide people, and escalators have replaced elevators to encourage motion and interaction. Tables and chairs are made partly of aluminum, and murals and collages give each kitchen a unique look. Best of all, suddenly everyone has a room with a view. No workstation is more than forty-five feet from the nearly twelve-foot-high windows looking out on the Allegheny River and the Pittsburgh skyline. Centrally located coffee areas and cafeterias on each floor are magnets for

collaboration. And everyone, including Chairman O'Neill, gets the same nine-by-nine-foot open office.

The result wasn't merely that Alcoa won a *BusinessWeek/Architectural Record* Award, sponsored by the American Institute of Architects. O'Neill told the *New York Times* that the new offices have helped them to "beat the liver out of the competition."

But you don't have to be a Fortune 500 company to realize that space matters. We have faced similar issues in trying to make the best use of our San Francisco office. The site has natural advantages and possible drawbacks. Located near the foot of the Bay Bridge in an old maritime pier, the office was designed to include a spacious deck and a glorious sixty-foot-long view of San Francisco Bay and the bridge overhead. But one-third of that view was lost because a wall had been built without windows to save a few thousand dollars.

Eventually, we knocked out the windowless wall, opening up that space so that three roll-up Plexiglas garage doors could span the front. Moving the manager's office away from the new window allowed the sunny waterfront area known as the Lookout to be used entirely for shared purposes. Now lunch tables line the front, with a kitchen at one end and a periodicals library at the other; Plexiglas cabinets display dozens of current magazines (with back issues tucked inside). Workers can eat lunch and peruse a magazine on the couch or one of the rollable egg-shaped chairs. Every Monday a different member of the office hosts a lunch in the space. Large white foam-core panels break the space up from the rest of the office, so that the Lookout can be used to prototype projects or talk about company issues.

The space most companies might have doled out to executives is now an ideal place to greet or entertain clients, eat lunch, collaborate or read or research something from the library. Most corporate libraries are an afterthought, a lost opportunity when you consider the wealth of inspiration and information they might provide. The IDEO San Francisco office has made its library an integral part of the work space. The periodicals library is just part of a massive collection that runs the en-

tire length of the building and includes extensive reference materials and books. Laser printers positioned within the space encourage workers to regularly visit and use the library.

I think our San Francisco office is a good example of how you can subtly alter space to support innovation, make resources more readily available, and open views up for everyone. Try looking around your company. Might moving the executive offices make a change? Could making the library a part of the workplace help people find what they need faster? Should that great view go to the CEO or the company lunchroom?

TELL STORIES

Show me a company going places and I'll show you a story worth telling.

If your company has been doing some neat stuff or has got big plans for the future, it's probably got a place for show-and-tell. Hewlett-Packard, for example, is proud of its museum. The company employs an official, full-time corporate librarian who lets visitors see and sometimes touch the company's wide range of products. The librarian also publishes the oral histories of retired company executives.

Now, I'm glad they're preserving the legacy of this great company, but it's a bit like resting on your laurels. Wouldn't it be better if instead of a museum you displayed current projects that teams could be proud of and look to on a daily basis for inspiration? What about the latest prototypes, materials, and mechanisms? Wouldn't that be a greater influence on future innovation than canonizing the past? How about aspiring to a corporate "Exploratorium" instead of a museum?

We believe in letting the world know we've done some pretty cool stuff—without stifling that creativity in a museum. If your company

has done something impressive that tells a story, try to put it front and center in a place that makes your workers proud—and lets clients and visitors see what you do.

The lobby at 660 High Street is our best storytelling entrance. There's the sleek soapbox car that won the intensely competitive local race called the Sand Hill Challenge. On each side of the broad lobby, shelves overflow with prototypes and final products—everything from the Heartstream defibrillator to the Palm V, Crest Neat Squeeze, Pepsi Twist 'n Go cup, and the SoftBook electronic book. Nothing is placed behind glass—it's all there for us to handle and show visitors the nuances of innovation and design.

Each of our offices has some version of a storytelling entrance. At some, the story winds through the building and into the individual "neighborhoods." Storytelling is best when it comes from the heart. We encourage studios to display their best work, but we don't bog down the process with corporate rules. There's no official "storytelling" policy. Bureaucracy would just kill the inspiration. The best storytelling happens when the team itself is proud and wants the world to know what it's accomplished.

Our machine shop used to be one of our best secrets. Unlike many firms, we've employed and cultivated top model makers and machinists in-house. It's a competitive advantage that our clients have come to rely on. But for years, the shop never displayed the imaginative solutions they'd helped dream up for what often seemed intractable problems. One day the head of IDEO's shop asked why they weren't on the visitor tour. They, too, wanted to strut their stuff, as well they should. Today the shop has a compelling display that features everything from injection-molding equipment to stereolithography parts. Morale has never been higher.

Storytelling is usually inspired by one or two individuals. Just as too many cooks can spoil the broth, too much technology can kill a good story. I believe in a healthy skepticism toward capturing a company's achievements in formal knowledge bases or databases. There's a cer-

tain art and serendipity to great storytelling. It should be fun. It should be about letting workers express themselves.

MAKE YOUR JUNK SING

At many companies Dennis Boyle's cluttered corner might be considered a liability. Truth is, Dennis used to wonder if there was something wrong with him. He'd been collecting mechanical parts, odd materials, and other quirky artifacts since he was a boy.

As the head of one of our most celebrated studios, Dennis became known as the keeper of the files, the stuff, the slides, the boxes. Circuits, fasteners, exotic plastics, unusual glass, oddly molded parts—you name it, Dennis had it. As a Stanford lecturer, Dennis often lugged around a cardboard box to show his students. He called it the "magic box" and told his students that engineers should make a practice of collecting interesting things that might one day prove useful. In essence, Dennis was celebrating a physical version of what writers and lots of people do with a diary, writing down interesting events and insights. Often, to spark a creative brainstorm, Dennis would bring a few choice items from his magic box.

When Dennis became a studio head, he searched for similar ideas to inspire his team. What about the magic box? he thought. He knew that plenty of other engineers and designers had impressive collections of gadgets and materials, but unlike his, theirs often weren't accessible. He decided to invent a model to make the gadget collection a part of our process. No corporate dictum was issued, no meetings called to poll the studio heads. Dennis simply grabbed a Stanford intern, pointed to his boxes, and asked him to come up with some sort of inexpensive, interactive display.

There was some gentle resistance. My brother David kidded Dennis about his obsession when he kept bringing up his project at our firm's

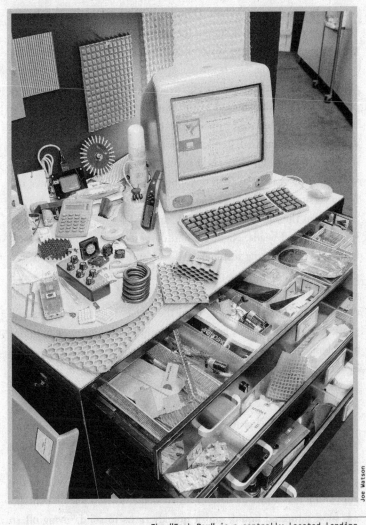

Joe Watson

The "Tech Box" is a centrally located lending
Library of innovation elements.

roundtable meetings. When Dennis encouraged other celebrated IDEO collectors to pitch in their masterpieces, he was met with skepticism. Some figured they'd never see their prized possessions again.

But ultimately the "Tech Box," as it became known, was a runaway success. With a little proselytizing, nearly everyone started generously offering up their valuables. A five-drawer flat-file cabinet, designed for blueprints and interlaced with bento-box-style partitions, displayed the eclectic collection. Each item was neatly tagged. Set smack in the middle of our 660 studio office, next to the printers and copiers and DC-3 wing meeting space, the Tech Box soon became our process center, our living version of the lobby shrine and museum.

This was the Montessori approach, and just as kids learn to count by moving colored beads, engineers and designers came to this physical and mental wellspring to touch and feel and see. The Tech Box became our corporate spark plug. Stumped on a project? Rifle through some drawers at the Tech Box. Need inspiration for a project or thorny design issue? Grab something from the box to toss on the table and spark new ideas. The box holds tiny switches and minuscule batteries and plenty of exotic materials—everything from Aerogel, an amazingly light translucent substance that feels like a solid form of smoke, to Mylar, the golden satellite insulator, to Kevlar, the material that stops bullets cold.

Today clients and visitors flock to the Tech Box to see—and more important, *hold*—the miracles within each drawer. Stir a cup of boiling water with a thin, long copper heat pipe, and the end of the tube instantly burns your finger. Drop the Nitinol "memory metal" wire in the cup, and it reassumes its natural coiled shape. Twist a packet containing supersaturated liquid, and it quickly turns to solid, giving off so much heat that it nearly burns your hands. Rubber balls that don't bounce, opaque materials that become translucent with the flip of a switch—the only common denominator in the Tech Box is surprise. "This is more about serendipity," says Dennis. "Making a lateral connection between slightly odd things that might give us a different perspective."

The Tech Box is our constantly evolving eclectic LegoLand. Everyone aspires to be part of this living show-and-tell. One day I contributed a Pack-A-Cino—a cappuccino with its own heat source. It's a novelty, but Dennis feels it might someday work in redesigning some piece of sports or emergency medical equipment that needs to be highly portable.

The Tech Box straddles process and architecture. It's both a mindset and a physical statement. It's a heck of a lot more valuable to our work and culture than the typical corporate art purchase—some twisted hunk of steel welded by an abstract sculptor. The Tech Box is our own private beach of imagination, a place to let the sand run through our toes and catch a few waves.

Today the Tech Box has been adopted at all our IDEO offices worldwide. Curators continuously seek out new technologies to collect, evaluate, and distribute around the firm. We've now cataloged the two-hundred-plus items in an Intranet database that offers some background on each technology, as well as any practical uses we've figured out for it on a project. This Website is accessible from everybody's desk, or from an iMac on top of the Tech Box itself (next to the human-powered windup radio and the martini glass that never spills).

Dennis's pet project has blossomed into a strategic IDEO advantage. Partners like Steelcase subscribe to the Tech Box (to receive actual copies of the items and updates), and we're considering offering it to select long-term clients. More important, the Tech Box has become a launching point for many projects and helped to improve our creative process. So many workers submitted technologies that Chris Kurjan, one of the Tech Box curators, had a tough time navigating the boxes of rejects engulfing her work space. IDEOers would finish a project with cool samples and materials and be loath to throw them away.

That wealth of great stuff inspired her to create a manufacturing processes display. IDEOers could come to the display to learn how to make things out of plastic and metal—say, the case parts for a remote control, the bezel for a medical instrument panel, or stunning lithographic finishes. It wasn't long before Tech Box II, the Manufacturing

Process Story, was playing in IDEO offices across from the original Tech Box.

Dennis Boyle's once underappreciated cardboard boxes have taken on a life of their own. When space celebrates process and teamwork, you know you've got something you can grow with.

For many companies, space is among the least considered, most overlooked tools in the innovation toolbox. The hundreds of companies I have seen firsthand spend a *lot* of time on structure and strategy and systems and staff. When it comes to space, however, they think of it as a mere utility. Space is treated as a threshold variable, made just good enough, except for an expensive flourish in the lobby or the main conference room. At IDEO we believe space matters, and I hope you will too. We all know the legend that innovation starts in a garage, but sooner or later we all grow up and need a place to work. No matter how electronic your business may be, there's a physical element too. Being innovative and successful is more than hiring the right people and buying the best technology. You've got to create a culture where space matters.

istory teaches that innovation does not come about by central planning. If it did, Silicon Valley would be nearer to Moscow than to San Francisco. Working on thousands of projects has taught us a simple but critical element that every team or company should come to expect: the unexpected.

Chance offers insights you didn't anticipate. It's a well-accepted truth that inventions and discoveries often result from random accidents or experiments that went awry.

How can you capitalize on that phenomenon? Well, you can start by expecting the unexpected, being open to surprises from sources within and outside your organization.

Try approaching projects with humility and the knowledge that answers may come from places you least suspect. We call this "looking cross-eyed" and "cross-pollination." It's quite liberating and powerful. If you expect to find answers from unusual places, it's far more likely to happen. In nature, we know that cross-pollination leads to superior strains of plants. It's the same with products and services. Launch a project with the assumption that cross-pollination may help you to innovate, and you're more likely to be ready to take the leaps of creativity necessary for innovation.

Now that we're midstream in the innovation process, I think it's appropriate to issue a disclaimer. We don't know it all. Neither does your boss or even the CEO. Accidents happen. Here follow some of my experiences and insights about how it is that serendipity plays a critical role in innovation—and what you can do about this puzzling variable.

BALANCING UNPREDICTABILITY

Growing up near Akron, I had "Ohio history" classes that included the story of Procter & Gamble's fabulously successful Ivory soap. It began with a blunder: an ordinary worker who went to lunch and accidentally left a mixing machine running with a batch of soap inside. When he came back, the mix had been whipped to a froth, with soap so light that it floated. By chance, the new floating Ivory proved to be both convenient and popular, marketed by P&G as "99 and 44 one-hundredths percent pure." A factory worker stumbled onto a new way of mixing soap, but it was P&G's marketing group that seized the opportunity to create one of the most successful packaged goods of the twentieth century.

Chance has played a role in all kinds of breakthroughs, from science to technology and business. Velcro started when a Swiss mountaineer returned from a hike covered with prickly cockleburs. A little observation under the microscope yielded the reason—nature's tiny hooks, perfect for grabbing a ride in the feathers or fur of passing creatures. *Voilà!* Soon the French were making Velcro's artificial hooks and loops that stick and easily separate. People had been plagued with prickly burrs from the beginning of recorded history. It took an intellectually curious person with access to the right tools to turn the idea into a breakthrough material. Stop and think for a moment. Are you proactively hiring the intellectually curious? If so, what tools do they need to help them along?

Saccharin was discovered in 1879 when a research fellow at Johns Hopkins University found his bread extra sweet one night and figured that something from the lab must have followed him home. Incredibly, he set about to tasting nearly everything in his lab—and lived to find o-benzoic sulfimide—saccharin by any other name. Chance not only favors the trained mind, but in this case, it favored a man brave enough to lick everything in sight till he could trace the source of his discovery.

UNEXPECTED UPS *AND DOWNS*

Successful companies have the agility to turn small failures or anomalies like these into big successes. Unfortunately, the reverse can also occur. Even after two decades of watching our clients' products rise and fall, we occasionally bet on the wrong horse. Sometimes it seems that guessing which product will fly is a little like picking the classmate most likely to succeed. Two that really fooled me were Dynabook and Momenta.

I wasn't the only one who thought the Dynabook laptop computer looked like a winner. It was one of our sleekest portable designs, a black cast-magnesium computer that looked ahead of its time. Dynabook Technologies even had backing from blue-chip venture capitalists at Kleiner Perkins who had a stellar track record of sensing "the right stuff." But there was one thing Dynabook hadn't planned for—Intel coming up with a new chip. Just when Intel 386 chips were becoming available, Dynabook introduced its 286 machine. All the good looks and well-integrated features in the world couldn't carry an old chip around the block.

In the same way, Momenta was one of the handful of IDEO clients I was ready to invest in from the very beginning. Headed by the charismatic Kamran Elahian, the company seemed to have both history and money on its side. Kamran had a great track record and so much sway with venture capitalists that he was turning away money. And the company's marketing campaign was smart. It focused on what Momenta's pen-based computer would do for real people—along the lines of our "verbs not nouns" approach—and had a captivating "Thousand Days to Greatness" program to motivate the internal development team. Momenta spent nearly $3 million in marketing on the day of the product's unveiling alone to make sure that the world noticed its product launch.

We'd already worked on just about every other hope-springs-eternal pen-based gizmo in the Valley—the GRiDPad, the PalmPad, the Go PenPoint machine and its spin-off, the EO. But Momenta was a powerful computer and—to me—seemed to have an aura of success about

it from the very beginning. In retrospect, maybe it was trying to do too much; its large, high-resolution touch screen was expensive; its hand-writing recognition wasn't quite there yet. This much I know for certain: Momenta went on to become one of the most glorious flameouts in Silicon Valley history.

GOOD TIMING DOESN'T HURT

Over the years I've noticed some patterns. Products that become hits seem to enjoy a balance of features, price, and that often elusive element of timing. And lightning can strike for what on the surface seems a humble product. We've designed plenty of computer monitors over the years, for instance, but our most successful monitor is one of our most understated. The "Simply Samsung" SyncMaster monitors have sold in the millions of units, more than any other monitors we've designed. It may not always make headlines, but good design at an affordable price sells.

Some products seem to be blessed with perfect timing and sizing. The Polaroid I-Zone camera was one of our more ambitious projects (we did extensive observations with kids to find out what they found "cool"). Still, I wasn't sure that the postage-stamp-sized photo stickers that had been so successful in Japan would catch on in the United States. Yet in the last few months of 1999, it became the best-selling camera in America. Why? Though there's long been strong demand for Polaroid cameras, the market was partly constrained by the fact that you needed a fairly bulky camera to develop a reasonable-sized print. But since the I-Zone prints tiny pictures, the camera itself can be kid-sized. And it turned out that the smaller pictures provided a tremendous opportunity—they're ideal for sticking inside a school locker or on a notebook. In a nutshell, the I-Zone translated the instant gratifi-

cation of the adult-sized original to kids, for whom instant gratification is a way of life.

NIMBLE RESPONSE

Just as we often can't predict a product's success, companies can't always divine what feature or use will catch the public's imagination. For that reason, companies need to be in touch with what "quirky" uses consumers have thought up for their products, and be ready to restructure their marketing accordingly. Kleenex, for example, was intended to be little more than a niche product, essentially a disposable cold cream remover. Only when the public started wiping their noses with the paper hankies did the Kimberly-Clark Corporation see the potential and begin to totally reposition its product as a disposable handkerchief.

With the benefit of 20/20 hindsight, it's clear to us what Kleenex is all about. But just how a new product might be used in the marketplace is often nearly impossible to divine. DuPont originally developed Kevlar to replace the steel wire in radial tires. But it was the aircraft industry that first responded to the exotic fiber, recognizing that its lightweight strength could save millions in reduced fuel costs. Police forces discovered that Kevlar vests weighing less than three pounds could stop a .357 Magnum bullet, and the U.S. Army found it a safe and lighter replacement for steel helmets. Canoes, kayaks, and powerboats rode the Kevlar wave, then came ropes and cables, and soon Kevlar was everywhere, replacing everything from nylon to steel in car and machinery hoses and belts, protecting workers against fire, and spreading to even more exotic uses like lightweight parachutes and aircraft ejection seats.

Eventually, Kevlar *tires* did take hold, but this amazingly strong and

resilient material didn't catch on until people started experimenting with it.

Zelco, the innovative company that brought us the "Itty Bitty Book Light," made a water-resistant clock with a rope on it that you hang in the shower. The product chugged along with unremarkable sales until one day demand shot through the roof. The company had no idea what hit them until someone clued them in that a popular rap band had appeared on MTV wearing Zelco's wall clocks around their necks. The spike in demand may have taken them by surprise, but nimble entrepreneur Noel Zeller was on the phone the same day, ramping up production and revamping his marketing for the product.

Likewise, the venerable Hush Puppies shoe, introduced in 1958, was nearly discontinued for lack of sales in the early nineties. Suddenly, spurred by grassroots consumer interest, Hush Puppies became the darling of the fashion industry, in demand at New York City resale shops, worn at hip-hop dance clubs, and even shown on designer runways. Wolverine Worldwide, who makes the shoes, helped the momentum along with an aggressive new branding program. In 1996 Hush Puppies won the prize for best accessory from the Council of Fashion Designers. Wolverine has since reported record earnings and is now the largest U.S. manufacturer of footwear. A sudden—and unexpected—turnaround for a product on the verge of being discontinued. Being alert to fashion trends helped the company take advantage of its windfall.

LOOKING CROSS-EYED

Over the years IDEO has learned that ideas for new products and services often come from left field. We've been aided by the fact that we work in diverse fields—high tech, medical, office furniture, e-commerce, transportation, and architecture, to name a few. It's a fact of life. If you see and take part in more innovation in disparate indus-

tries, your innovation vocabulary is bound to increase along with the opportunity for cross-pollination. But any company can take advantage of this factor. The first step is to learn how to look beyond your own backyard. At IDEO we often begin with questions. Could something in our latest Internet appliances be transferable to one of the medical projects? Might we be able to work some of our new computer interface ideas into a consumer product?

I've been in dozens of exciting brainstorms, and what always amazes me is how many good ideas drift over the fence. Our *Nightline* shopping cart casters sprang from an office furniture prototype, and the first wheels for the Sand Hill Challenge soapbox race came from high-tech wheelchairs. When IDEOers talk about thinking outside of the box, we're not talking about just going beyond what's typically done in a particular field. We often begin by wearing the glasses of totally different disciplines.

Take beach chairs. Several years back, somebody came to us with the sort of strong observation that often inspires good products.

Daniels & Daniels

The idea for this Sun Tracker Beach Chair came
from a tilt-and-swivel computer monitor.

Wouldn't it be nice to have a beach chair where you didn't have to get up and turn the chair every hour or so to follow the sun? So where'd the solution come from? During a brainstorming session, someone brought in a computer monitor with a ball-and-socket mechanism used for tilting and swiveling the display. We adapted that computer concept (itself an adaptation) for the beach. The result was a "sun tracker" beach chair with a rounded base that "scrunches" down into the sand, which acts as the pivot.

Another chair we worked on—this one for the office instead of the beach—had a tilt mechanism underneath with a troublesome metal coil spring. We solved the problem by creating a cylindrical adjustment handle—made by bonding rubber to metal—which could double as the compression spring itself. The safer and more comfortable adjustment mechanism came about partly because two IDEO engineers had experience in the automotive industry, where bonding rubber to metal is commonplace.

Watch people struggle with existing products in new environments, and you'll likely sense an opportunity for cross-pollination. One way to open your mind to this possibility is to ponder how products made for only *one* use are adapted by clever people or companies for different uses.

There's the milk crate, that delightfully versatile container used for everything from filing to furniture to clothes hampers. Milk companies *hated* how popular those crates were, and as far as I know, no dairy ever turned the vice into a virtue by selling them on the open market. Every start-up worth its stock options knows that doors make for good cheap desks. Duct tape, originally intended for sealing galvanized heating ducts, has become one of the most versatile items in a handyman's toolbox. The golf cart, of course, is used by Hollywood studios and corporations around the world for quiet, on-site transportation. Then there's the case of Breathe Right nasal strips, a device made to help people stop snoring. San Francisco 49ers receiver Jerry Rice started wearing this Band-Aid-like contraption on his nose to improve his air intake, and sales suddenly soared—for athletes as well as snorers. And, of course, there's the personal computer. Highly complex problems

used to be tackled by supercomputers. Then some very smart computer scientists realized you could take ordinary personal computers and channel their computing might through parallel processing. Today you can achieve much the same level of computing power by activating distant machines on the Internet.

Not too many years ago the only mass-produced sports goggles were those made for downhill skiing. Motocross riders, who know something about getting through a rough course, patched up the goggles as best as they could. Since mud would quickly cake up on the lenses, they jury-rigged traditional ski goggles to hold half a dozen clear lenses.

Daniels & Daniels

Smith's clear "Roll-Offs," helped adapt ski goggle designs to the muddy world of motocross.

During a race they'd peel off the muddy lenses one by one to temporarily regain an unobstructed view. Ski goggle leader Smith Sports Optics, knew there had to be a better model. They found inspiration in rolls of 35mm film, and created a clear Mylar roll. Pull a cord on the "Roll-off" cartridge, and you'd get another fresh strip of Mylar, and a clear stripe across your goggles. Roll-Offs held enough clear film to outlast the dirtiest ride.

EXERCISING YOUR CROSS-POLLINATION MUSCLES

The Tech Box, of course, is our conscious exercise in cross-pollination. Bringing quirky Tech Box items to brainstormers often sparks people to think beyond the norm. In the previous chapter I mentioned a thin copper heat pipe that burns your fingers if you dip one end into boiling water. It has its roots in the aerospace industry, where it's used for cooling critical parts. IDEO's brainstorm-fueled idea was to adapt it to cool off those hot microprocessor chips in mobile computers. It's a particularly elegant engineering solution. We've all experienced the drawbacks of fans, the most common method of dissipating heat in desktop computers—they're bulky, they suck up power, and they're noisy. But these tiny copper tubes silently carry away the heat.

When Apple came to Dennis Boyle looking for a nonmechanical method to eject PC cards from its PowerBook, he went to our Tech Box, a natural starting place, where he found Nitinol, that magical material that returns to its original shape when heated. He worked up a couple of quick prototypes, and sure enough, figured out that the "shape memory" alloy could be used as a kind of metal "muscle." A small amount of battery power could heat a tiny Nitinol wire, make it contract, pull open a latch, release a spring—and shoot out the card. Light, small, and highly reliable, Nitinol has lots of benefits. I expect it will find its way into other consumer products.

. Cross-pollination can be surprisingly low-tech as well. Sorbothane, that wonderfully springy, elastic material found in better running shoes, is another Tech Box resident. Where did we use it? First, we packed it in the Plus Passport removable disk drive, a device engineered to withstand shock. Later we used tiny amounts of the material to protect delicate components inside portable computers and other mobile devices.

How important can a little Sorbothane be? Well, we recently used it inside the set-top box for TiVo's personal TV service, a system that digitizes and stores TV broadcasts so you can pause or even rewind the show you're watching. In order to accomplish this magic, the TiVo machine houses two large disk drives that are spinning all the time. Critics found the competitor's product irritatingly noisy, but not TiVo. The difference in our design? We packed the TiVo's drive with a few cents' worth of Sorbothane to dampen vibration and cut the noise.

SEVEN PLANTING TIPS

What exactly is cross-pollination? Is it something that strikes you out of the blue? Can you plan for it? The truth is there's a yin and yang to cross-pollination. There's the active, aggressive side: learning about new processes, methods, and technologies. And there's the passive, accepting side: making time or creating a place for new ideas. The best cross-pollinators encourage both of these practices. Here are a few of the ways we try to make cross-pollination an integral part of the workplace.

1. SUBSCRIBE AND SURF

In your free time, browse magazines (and books) and don't forget to surf the Internet. We call this "idea wading"—so the farther afield the

better. We've got subscriptions to more than a hundred magazines, everything from *Abitare* to *Zoom,* and we get a kick out of checking out the latest offbeat website. Tom Peters says he routinely picks up as many as fifteen magazines from the airport newsstands and then spends the entire flight tearing out pages from interesting and offbeat articles. Around IDEO café tables, employees routinely flip through magazines, exposing themselves to the latest ideas, shapes, and advertisements. Bike magazines can be particularly inspirational, since bicycle companies constantly experiment with new materials. Browse the aisles at your favorite bookstore and make a practice of looking at sections you've never checked out before. You'd be surprised at how much you can learn about the latest trends by scanning titles and book jackets.

2. PLAY DIRECTOR

Take on the perspective of a film director. Break the world down into scenes and become expert at watching people perform even the smallest tasks. Like pumping gas. Flying a kite. Or carrying stuff to the beach. Watching events outside your normal sphere can heighten your observation skills. Take in a cricket match. Watch workers at a loading dock. Visit a factory. Or check out the FedEx man in his truck during the late afternoon crunch.

3. HOLD AN OPEN HOUSE

Periodically hold open houses to spread the best disciplines of your company far and wide. Keep it casual, displaying a few crude prototypes or perhaps a quick posterboard describing what you're working on. Photos, sketches, or some simple graphics work well. Combine it with some food to get people talking. Encourage comments and ideas, and you'll be surprised at how much you'll learn talking to visitors from other departments.

4. INSPIRE ADVOCATES

People talk about racial and cultural diversity. Mental diversity can be just as important. We all need individuals who celebrate different viewpoints. Rickson Sun has become our de facto green person, skilled in reducing packaging or energy requirements. Jane Fulton Suri's years of careful observations have made her a great representative for the elderly. Brendan Boyle's prodigious toy inventing has made him an expert on kids. We've got half a dozen other advocates. They've become part of our fabric. They make us all a little more conscious of critical constituencies and considerations in every new project.

5. HIRE OUTSIDERS

Fresh blood invigorates a company and introduces new ideas. Hire people just slightly off center, and you'll be pleasantly surprised. Dennis Boyle and Tim Billing, two IDEO leaders, came from General Motors, which might seem odd, since at the time we'd done no work for the car industry. And, of course, there's Jim Yurchenco, the first IDEO hire, and one of our most solid technical performers over the years. Yurchenco had no formal training in engineering and spent most of his college and graduate years soldering sculptures out of metal.

6. CHANGE HATS

We've already talked about the importance of observations. One of our very first clients was Telesensory Systems, creating reading machines for the blind. Team members were taken to their own house, blindfolded, and told to try making a peanut butter and jelly sandwich. Fumbling around without the benefit of sight, some couldn't find the jars, others the bread. Spreading on the peanut butter was tough. They gained a small but tangible appreciation of the challenges of blindness.

The best new products are like shoes someone else has broken in. Smart folks have already anticipated likely problems and opportuni-

ties. The difficulty most companies have when they try to do this is that they have trouble getting out of their own shoes.

Great products or services tell stories. It's a conscious narrative process. We often work up compelling characters to represent some possible users of new products. We might sketch them, give them names and families—background stories, in movie business lingo. The next time you start a new project, try dreaming up a few of these archetypes to help you see a little more broadly.

7. CROSS-TRAIN

Some athletes improve their performance in an event by training in multiple sports. We, too, can benefit from little drills or processes that we pick up from other businesses. Even the best companies can suffer from some overlooked weak internal business system. When my brother David joined NCR Corporation twenty-five years ago to work on early ATM machines, he found out that the company had a hopelessly disorganized scheme for tracking product drawings and documentation. But David had learned a great system during his brief tenure at Boeing, so he introduced a similar approach at NCR.

This principle, of course, is exactly why businesses have boards of directors with executives from widely different fields. Sometimes we focus too much on "the competition." IDEOers spend as much time learning from noncompetitive industries. Simple solutions to many problems are often just down the street—waiting for you to find them.

BARRIER JUMPING

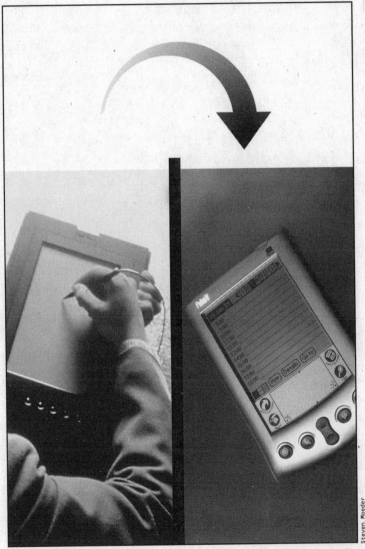

Steven Moeder

It happens on almost every project we work on. You observe, brainstorm, prototype—and still you occasionally stumble.

You're facing a barrier. The barrier might be external—cultural resistance to a new product or technology. Or it might be an ingrained ritual. How people perceive and use products often handicaps innovations. Companies get this wrong more than almost anything else. What people will accept or reject is a fast-moving target. And to make it even harder, you've usually got to predict this a couple of years out in the future. If that's not bad enough, there's the traditional barrier of patents.

The first step is anticipating and recognizing these barriers. The biggest barrier to innovation, of course, is a company's mind-set. We've assisted and watched some of our partners leap what seemed to be pretty high barriers; though in truth often the best path is to sidestep the obstacle. We've learned a lot about the little "handshakes" you can make with the people who use your products. We've learned that success often lies in embracing the spirit of the best of rituals. And we know that clever, coordinated effort—good old evangelism by any other name—can often get you over that hurdle when there's no way around it.

Barriers often stump even the best innovators. They wall in your imagination like a prison yard. Several years back, for example, my brother David figured the latest innovation was asking too much of people. He was being given a personal demonstration of a radically new mobile computing interface. Though impressed, he believed the device would face an uphill battle. People won't take the time to relearn how they use a computer, David argued. Why would someone want to learn a whole new way to *write*?

Over the years, David has been pitched thousands of inventions and innovations, everything from wearable computers to self-closing toilet seats. If there's one thing he's learned, it's that you should never underestimate the barriers to people adopting a genuinely new product.

So David seemed the ideal person to gauge the success of this particular innovation. He knows his computer interfaces and even teaches a course on the subject at Stanford. IDEO made the first commercial mouse, plenty of trackballs, joysticks, and virtual reality headsets, as well as more than a few radical input devices that never found a home.

But the simple shorthand called Graffiti offered benefits that far outweighed the time and effort it took to learn. And the technology was in the right hands. The man giving David the personal demonstration in January 1995 was none other than Jeff Hawkins, a veteran of GRiD Systems, makers of the original laptop computer. And Hawkins had tasted enough failure in the mobile world to make him battle-ready.

David was right that Hawkins faced some tough hurdles. And not only with his novel interface. Half a dozen major efforts at PDAs (personal digital assistants) had flopped in the previous few years, taking with them a billion dollars of capital. But the real and perceived value of Hawkins's new technology, plus a few smart bridges to make the leap easier (which I'll get to in a minute), made the Palm Pilot a runaway hit.

Of course, Jeff Hawkins and the Cinderella story of the Palm Pilot is the exception to the rule. Why are the barriers to adoption of consumer products so high that seemingly intelligently conceived and well-marketed products stumble? Is there a way to bridge the gap?

Entrepreneurs like Hawkins tend to have an intuitive sense of the barriers to adoption but most of us could use a little practice in recognizing and responding to potential obstacles. A good place to start is to look overseas.

WATCHING THE BORDER

Oceans and cultural divides can be big barriers to change. Not surprisingly, many high-tech consumer products like cell phones are embraced by diverse cultures, largely because there's no history or

tradition attached to them. Meanwhile, products with a long heritage like hand tools or furniture or kitchen utensils can be problematic. Cars and personal stereos have proved to be fairly international. But you can be surprised by what is not. Take vacuum cleaners, for instance.

Our research discovered that Japanese consumers prefer their vacuum cleaners quieter and less powerful with smaller motors than the American vacuums. Japanese apartments and homes are much smaller than American ones and the walls are thinner. Some Japanese shoppers will pay for "fuzzy logic" features that actually turn down their vacuum cleaner's power if they detect that a rug isn't too dirty. Japanese customers also are quite willing to pay extra for insulation to dampen the noise of their vacuums.

Americans, on the other hand, want their vacuums *loud*. We may not ask for them that way, but we associate the noise with a powerful motor and more suction. "If it ain't loud, it ain't working" seems to be the thinking. Our advertisements tout amperage the way hot rod magazines brag about horsepower. A feature that slows down the motor? Why would I want that? Once you know that the underlying desires of the two consumer groups are so different, it's easy to recognize the difficulty in trying to make a "global" vacuum cleaner.

Adoption curves vary tremendously from culture to culture and from product to product. Think of drink cup holders in cars. Who could argue with the benefits? By quickly integrating cup holders into their cars, Lexus and other automakers adeptly mastered one of those small but telling features that can make such a difference. Yet some European manufacturers were slow to embrace cup holders. Why? Perhaps because Europeans spend far less time commuting, drive shorter distances, and seldom pick up beverages at the drive-through window. But the mind-set of some European sports car makers also played a part. Some assumed their cars were such a pleasure to drive that no one would possibly want to compromise their performance or spoil the experience by taking one hand off the wheel to hold a cup of coffee.

The adoption of technology can be equally puzzling. Who would have predicted twenty years ago that the world's highest concentration of cell phones would be in Finland? What were some of the bridges for this tiny northern European country? First there were the commercial and national incentives of having world-leading cellular technology. Finland also has some of the lowest mobile phone rates in the world as well as the world's first global system for mobile communications. Nor did it hurt that Finland has long had a tradition of communication expertise. Early in the century, it boasted more than eight hundred local phone companies.

Yet there are technologies that inexplicably lie dormant for years before catching fire. Why did several decades pass before the fax machine invented in the early 1900s, took off? Since early faxes were expensive and unreliable, they had few advantages over the telegraph, which was superior at shuttling words. In Japan, however, there were cultural incentives to innovation. The telegraph and other existing text-based systems couldn't easily handle the ideographic Japanese language. So the Japanese government launched a massive research effort and urged manufacturers to adopt a common communications standard. Not surprisingly, Japan became a world leader in digital faxes.

Clearly, myriad factors can influence the spread of innovation. The food industry offers rich examples of how tradition can defeat change. Take the story of Betty Crocker's rude awakening when it tried to peddle its moist cakes in England. They didn't fly. The English like dry cakes and cookies with their tea, often using their fingers. They didn't want a buttery, melt-in-your-mouth cake. Or condensed soup, for that matter, which also failed to catch on in Britain because consumers felt like they were getting half the amount for twice the price. In Japan thick American-style disposable diapers flopped because diapers there are changed more often and storage space is minimal.

You can't simply skate over cultural differences. We ignore tradition or widely held beliefs, however mistaken we might believe them to be, at our peril. There's often a holdover effect in which perception con-

tinues down a straight path, even after *reality* has turned a corner. Consider the case of cattle guards. To keep the beasts from wandering through a road, many ranchers install expensive steel cattle guards, slippery grates that the hoofed animals can't cross. But once the cattle have learned to be wary of the grates, all the ranchers have to do is stripe the road with white paint. The cows keep their distance.

Stupid cows, you're thinking. Actually, humans do the same thing.

Folks who grew up in the fifties or sixties tend to rinse off their dishes and put them in the dishwasher practically clean. That, despite the fact that our clients at KitchenAid tell us this ritual has long been made obsolete by newer, more effective machines. What's the lesson? Often you've got to work especially hard to overcome rituals or superstition.

STUBBORNLY UNMETRIC

To truly understand the extent to which people or nations can resist change and innovation, we need look no further than our own front door. Incredibly, the United States has become the last major holdout against converting to the undeniably better metric system of measurement; only Liberia and Myanmar keep us company. Americans for Customary Weights and Measures dubs the metric system a "European fad." One senator from North Dakota claimed our nation couldn't afford to convert. Richard Weinstein, NASA's manager of engineering studies, went so far as to tell the *New York Times* in 1996, "We got pretty well into the space station design using inches and pounds. We'll accommodate the metric system where we can, but we simply cannot start over. The expenses would be prohibitive."

Perhaps if NASA had mustered the courage to leap the imperial hurdle, they could have averted the $140-million Mars Climate Orbiter disaster. The team at the Jet Propulsion Laboratory in Pasadena used

the metric system to compute the force exerted by the probe's thrusters. But the folks at Lockheed were thinking imperial. The result was that the Mars Climate Orbiter entered the planet's harsh atmosphere about eighty kilometers low, and likely broiled to a crisp. Yet the October 1999 debacle didn't stir a debate over the inherent foolishness of a two-measurement system. The media generally referred to it as a math error.

Orbiter was unmanned, but other metric mishaps could cost lives. In 1983 a Boeing 767 flown by Air Canada ran out of fuel in midair because a manual fuel-level calculation had confused pounds with kilograms. Fortunately, the Air Canada pilot had experience flying glider planes, and, miraculously, no one was killed in the dramatic engineless landing.

Searching for examples of metric/imperial conversion problems, I didn't have to look very far. In the building right next door to me, we were recently designing a simple plastic belt clip for the Handspring

Steven Moeder

A metric mix-up: [left side] What the engineer wanted—a belt clip 12 centimeters high. [right side] What he got—a "belt clip" 12 inches high.

Visor. Sending the design out to a model shop for fabrication, we specified centimeters, but through miscommunication, the vendor was thinking inches. So what happened? You guessed it. The parts came back exactly 2.54 times too big, looking more like a backpack than a belt clip. I'll bet this sort of thing happens more often than most folks would like to admit. Engineers and machinists routinely say, "Take a mil off that surface." Trouble is, a "mil" can mean either a millimeter or a thousandth of an inch. No wonder confusion crops up.

Why hasn't the metric system taken hold in the United States? I think it's a simple lack of motivation. Nobody has provided a compelling reason to make the shift. Contrast that to the tradition-bound Brits, who created the imperial system by royal decree in 1215. Nearly eight centuries later, Britain boldly declared September 30, 1995, "M-day." Most of the nation's businesses were ordered to go metric, with offending shop owners fined thousands of pounds (in theory, anyway). The British laid clever bridges to relieve the fears of traditionalists, nostalgic exemptions for the beloved pint and mile, what the Department of Trade and Industry called "descriptive measures." Poets complained mightily, writing that phrases like the Merchant of Venice's pound of flesh—454 grams—just wouldn't be the same. But when all was said and done, the meter rose and the pound fell. Dragging our feet on innovation, America collectively assumes the cost of converting is too high. Yet someday—soon, I suspect—we'll decide that it costs too much *not* to convert.

SIDESTEPS

In 1990 GRiD Systems introduced the first pen-based portable computer that featured handwriting recognition. The product was a little chunky and sluggish, but reviewers saw GRiDPad's promise, and customers pined for a peppy, pint-sized computer. And so began the first

pen-based computer craze. Venture capitalists poured money into companies like Go Corporation, Apple accelerated development of the Newton, and John Sculley gave the phenomenon a name, the PDA, when he started rhapsodizing about his "personal digital assistant."

Jeff Hawkins, who had helped create the GRiDPad, left to create Palm Computing in January 1992, and threw everything into his first effort there, the Zoomer—complete with a minuscule keyboard, drivers for printers and fax machines, and handwriting recognition software. He learned the hard way that trying to be all things to all people for seven hundred dollars was a recipe for a consumer electronics disaster. Both the Newton and the Zoomer faded, followed by other short-lived products from Sharp, Hewlett-Packard, Go, EO, and Toshiba.

Remarkably, Hawkins got another chance. Laying the first bridge to adoption, he asked what had gone awry. Surveys of Zoomer customers gave him valuable insights. Following a long-honored tradition in the computer industry, manufacturers tried to cram all sorts of capabilities into the tiny devices. But most customers were more interested in something considerably less complicated. They were looking for a reasonable *complement* to their PCs, not necessarily a *replacement*.

He discovered another popular misconception as well. Hawkins and nearly every other vendor had been pounding their stylus against what seemed a fundamental dilemma. Handwriting recognition software has to recognize your handwriting. And since the software required to accomplish that was incredibly complex, PDAs had to be painfully slow. Or did they? Hawkins's research into cognition and neurobiology helped him sidestep this stumbling block. He'd learned about auto-associative memory, the human brain's ability to fill in pieces of the puzzle. Why couldn't people do part of the work?

Hawkins's radical idea was to get PDA users to do some of the heavy lifting. Slim down the software and hardware by off-loading some of the pattern recognition to humans. Create a language—Graffiti—that is simpler for the computer to understand.

Hawkins believed people would make the leap necessary to adopt Graffiti because his research had taught him that people genuinely

thirsted for powerful new tools. But he knew the learning curve couldn't be too high. He practiced endlessly with pen and paper, refining and simplifying his Graffiti alphabet before incorporating it into the Palm. Colleagues and friends were surprised at how easy it was to learn the streamlined, single-stroke letters. Hawkins's other big focus was the *size* of his PDA, and he found the ideal metaphor—the shirt pocket. Mindful of the Zoomer's excess, Hawkins challenged every feature's right to survival, and week by week the device got simpler. He developed a prototype in just a few months and in April 1996 launched the Palm Pilot. It took off like a rocket, gaining market acceptance at a faster rate than the color TV, the cell phone, and the personal computer.

The story of the Palm Pilot fits neatly into the S-curve theory that every technology or industry has a rate of inception, adoption, and decline. In the classic S-curve model, there's a long, slow development-and-acceptance period, until the product or service reaches an inflection point and suddenly surges. Since trees never grow to the sky, that growth curve inevitably tapers off into another flat part of the curve. Over the years, IDEO has ridden the S-curve like a roller coaster.

Innovations power products up the steep
part of the S-curve.

When we first started working in the computer field, machines were fixed to desks. Then, in the space of one explosive decade, the center of our universe leaped from luggable to portable and finally something you could hold in your hand. You could say that nearly all innovation programs are either about pushing a product past the inflection point or jumping to a new S-curve.

Our experience with pen-based computing has taught me that there's often a hidden barrier in the S-curve. Individual companies and entire industries often slow adoption rates by embracing faulty assumptions. To me, the more than half dozen highly funded pen-based efforts that preceded the Palm Pilot failed for essentially the same reason. Each was operating under a PC-replacement theory—the idea that a pen-based computer had to achieve 100 percent handwriting recognition. Since that was an enormously difficult technical hurdle, they all failed.

The real barrier to the initial PDAs wasn't the limits of software and hardware in a handheld box. It was the idea that the machine had to do nearly everything, that people couldn't and wouldn't want to learn. Personally, I find the simple Palm OS products compelling. I carry my Palm V everywhere, get a kick out of using the stylus, enjoy beaming contacts to other Palm users, and have found that I have a stronger connection to the Palm than I've had with any other computing device.

What's the lesson here? When the public is slow to accept a new technology, you need to reexamine your assumptions. Sometimes the wisest course you can take is to start with a modest step forward, like exploiting a niche. Early on, pen-based computing, for instance, made great sense for applications that didn't require lots of handwriting. Indeed, of all the pen-based machines we worked on before Palm, there's one that's still in service. Next time you rent from Hertz, you'll see it. It's the rugged little Norand Pen*Key computer carried by almost every Hertz bus driver.

There's another factor, of course, one that you won't find many engineers discussing. Thresholds of acceptance and rejection exist for

every new product and service. For example, while cell phones are starting to permeate American culture, it's still a more complex tale than you'd find among early adopters like Finland, where mobile phones are nearly ubiquitous and fully socially acceptable. In the U.S. there's still some lingering social resistance to cellular technology, even when it's not disrupting your dinner conversation. I frequently use my cell phone when rushing from one IDEO building to another, a span of several blocks. My unscientific study suggests that when I'm walking with my phone, motorists at the intersection are less patient, more likely to cut me off, and much more inclined to give me a glaring look. As I write, cities and states across America are contemplating bans on cell phones in restaurants, bars, and even on roads. Some people find them intrusive and disrespectful, especially when you're paying for an expensive ritual, like an exquisite meal in a fine restaurant. I'm not surprised that many people are in favor of a ban. Cell phones, as I've described before, are "verb" products, and thus move with us and create extraordinary opportunities and obstacles. It's an idea well worth keeping in mind, especially as all kinds of devices become mobile and tied to the Internet.

Social norms are notoriously difficult to predict and at times even tougher to change. There was a time when answering machines were considered rude. Then, in the space of just a few years, it suddenly became rude if you *didn't* have one. E-mail surveillance has long been feasible for corporations, but I doubt most workers embrace the technology (or for that matter, that it contributes to an innovative culture). People take privacy seriously. You can't overlook the intrusion effect. When we were once asked to do some work on a biometrics access device—the equivalent of a digital handprint—we learned a couple of hard lessons. Technically, you only need the rough outline of a hand. In fact, just the relative length of the fingers and the spacing of the knuckles can separate one hand from thousands. But we also found out that many people thought of handprints in the same way they thought of fingerprints. It's intrusive and makes them feel like crimi-

nals. That's why biometric access controls still raise some concerns, though it's a barrier that may be falling soon.

DEAL KILLERS

Silicon Valley marketing expert Regis McKenna talks about the FUD factor—fear, uncertainty, and doubt. FUD may not be totally rational, but it's very real. When we worked on the EV1 electric car charger, we discovered from our research that Americans are anxious about handling electric currents over 110 volts. The midlevel GM car charger was going to be 440 volts. Conventional wisdom suggested using a meaty three-pronged plug like the one on your electric dryer with metal-to-metal contacts. But getting people to embrace electric cars was already an uphill battle. The last thing GM wanted to do was to scare off potential drivers. So we helped them scrap the fork and design a meaty rubber handle that tapers to a paddlelike charger that neatly docks into the car. The energy is transferred by inductive current. The paddle not only *is* safe but also *looks* safe. Sometimes you need both.

It's a scenario we've seen countless times. More often than not with new products or services, you've got to acknowledge potential "deal killers." Reach out to the public, or at least meet them halfway. Think of early e-commerce, when people worried about their credit card numbers being stolen, and pundits declared we wouldn't shop on-line. Vendors responded with a series of strategies to diminish that risk and reluctance, features like the "secure connection" dialog box, displaying only the last few digits of your credit card number. And, of course, they also tossed in free shipping, coupons, and other e-centives.

Many new products flop because something critical is left out. Jake Burton Carpenter, founder and owner of Burton Snowboards, says innovation happens at his entrepreneurial company by looking for what *isn't* there. "We ask," he says, "what do people want? What's missing?"

A few years back, California inventor Tom Christianson came up with a fog-free shaving mirror for the shower. With high hopes, he showed the ingenious device—the trick is in circulating hot water behind the mirror—to senior management at Sharper Image. But the mirror required a pair of pliers to install, and Sharper Image doesn't sell "do-it-yourself" products. Deal killer? Not quite. Nimble Christianson figured out a way to accompany the mirror with inexpensive pliers. Sharper Image changed its mind, and shipped pliers with each mirror. Richard Thalheimer's ad copy read, "The installation is fast and straightforward, even if you aren't handy"—and the ShowerTek fog-free mirror was a success.

There are also the countless products that don't compellingly differentiate themselves from what's already on the market. Remember the Susan B. Anthony dollar? It was almost exactly the same size as a quarter. It's almost as if the U.S. Mint skipped the consumer research stage. Standardized coin sorters mistook the coins for quarters. Cashiers couldn't distinguish them from twenty-five-cent pieces, and people occasionally overpaid for items. The mint experienced something pretty close to a worst-case product launch scenario: a product that failed miserably and generated bad word of mouth.

So how do you recover from a humiliating disaster? The new head of the U.S. Mint took some key steps to make certain he didn't make the same mistake the second time around. The new Sacagawea coin dollar is substantially bigger and thicker than a quarter, and, just in case there's any confusion, it's tinted gold. It just might work.

PATENT PENDING

When launching a new product of any kind, what can you do when you're faced with what seems the ultimate barrier, a competitor who seems to have a whole product category locked up through a bevy of

patents? Is it possible to develop a product in the same category without getting sued?

It's a challenge IDEO has faced more than a few times. Many years ago Smith Corona came to us with an intriguing project. Remember the Dymo label makers that squeezed letters onto plastic strips that you'd peel off and stick on all your earthly possessions? Japanese manufacturers had created updated versions of the label maker, which they protected with more than forty patents. Smith Corona was interested in entering the market but worried about the formidable patent barrier.

One of the first steps in such a project is the obvious one. Somebody's got to start reading—in this case an intimidating stack of patents and more than six hundred claims. Rickson Sun, one of our most technical and scientifically curious engineers, led the project. He just started trying out all the existing label makers to see how they worked. He wasn't impressed. The patents mainly covered the process of opening a door and taking a cartridge out. Rickson thought the mechanisms were clunky and required too many steps. To replace the tape on one label maker, you had to flip it over on its back, unhook a latch, lift a door, wiggle the cartridge out, put in a new one, wiggle it into place, shut the door, and close the latch. Wouldn't it be easier if you could load the cartridge from the side and not have to flip it over? wondered Rickson.

When you're competing against an existing, established product, sometimes the best thing you can do is start from scratch. Rickson knew there had to be something easier for users. And that's how he came up with a simple push-button ejector. Push it and the spent cartridge would pop out. Indeed, the one-step-out, one-step-in motion worked so effortlessly that it was far easier to change label colors.

And it won a patent.

OPENING THE FLOODGATES

We've talked about some of the obstacles to better products and services. But there's another, tougher kind of barrier. It's the barrier in your mind, or in your office, and it can sap your strength like a virus.

We all know the feeling of being stuck. Writers call it writer's block. Stay blocked long enough and you'll really go into a tailspin.

What's the opposite of writer's block? This may sound a bit touchy-feely, but we've witnessed enough creativity that we've come to think of it as a stream of energy. We all recognize the euphoric moments when ideas and decisions come effortlessly. It's a kind of magic that loosens us up and lets us perform at our peak. Athletes call it being in "the zone" and psychologists call it "flow." But athletes who overtrain and students who cram too hard often find that zone of peak performance to be elusive. The trick is to find that delicate balance between complete choreography and pure inspiration. I've felt it when I tried to play golf. At the driving range, I seemed to momentarily master the mechanics of the game, but then I'd go out on the course and let one bad hole turn me into a shanking machine. And yet, every once in a while, the ball sometimes seemed to magically leap off my club when I was barely trying.

Creativity works much the same way. All the talent and intelligence in the world can hold you back rather than help to foster creativity if you are uptight. No matter what our God-given abilities and finely honed skills, we all can use a little help now and then. I think of it as trying to keep your bag of magical instruments in tune, working and playing at achieving a harmony that wards off the inevitable dips in creative juices.

In the stream of innovation, many companies make the mistake of building dams instead of doing everything possible to increase the flow. But when a culture is devoted to searching for breakthrough ideas, it's as if the river keeper has opened all the floodgates.

That said, droughts come and go. Creativity requires careful tending. Over the years at IDEO, we've collected a list of the dams, and the floodgates that increase the flow of creativity and breakthrough ideas.

Good pilots never take off before running through a checklist. I think we all benefit by periodically cleaning out our organizational and mental attic, seeing if our thinking and processes need some spring cleaning. Here's my own quick creativity checklist.

BARRIERS	BRIDGES
Hierarchy-Based Innovation and structure are like oil and water. Forcing ideas to start at the top or rigidly follow a vertical path through an organization tends to weigh down new projects. There are just too many obstacles.	**Merit-Based** If your company is truly willing to embrace ideas from any source, not only will innovations flourish more readily, but people will be more open about tossing their thoughts into the ring.
Bureaucracy If you have to fill out a standardized form or consult with a lawyer every time you start a project, pretty soon you'll just try fewer new things.	**Autonomy** If you're the master of your own destiny, you'll have the self-confidence to take risks. Yes, you'll occasionally stumble, but you'll keep reaching for new successes too.
Anonymous There are companies where nobody seems to notice or care. Places where you can cruise slowly up a predictable career path as long as you don't rock the boat. Places where playing it safe is the wisest course.	**Familiar** If you feel like friends and family, then someone will understand an occasional misstep. Good organizations make you feel comfortable enough to poke fun at each other—even at the boss—because you're among friends.
Clean Beware of clean desk policies or strict rules about customizing your space. Even a "tidy" organizational structure can stifle creativity. If	**Messy** My office may be messy but it's unmistakably mine, a personalized home base for the many hours I spend there. New York City is a

you keep laying down restrictions, you shouldn't be surprised when the project team has trouble "thinking outside the box."	good metaphor for this kind of stimulating messiness—a jumble of cultures, ideas and experiences that spawns great energy and creativity among its inhabitants.
Experts Expertise is great until it begins to shut you off from new learning. Many self-described experts, for example, talk more than they listen. Experts can inadvertently block an innovation by saying, "It's never been done that way."	**Tinkerers** A tinkerer is always tweaking things and ideas, trying to improve their work and themselves. Tinkerers ignore the status quo because they know they'll be trying something a little different tomorrow. They're great at launching projects and keeping them in motion.

SKILL SETS

Sometimes we shut ourselves off from opportunities because a skill seems too hard to learn. But with the right approach, all sorts of skills are within our reach. Take simple sketching, which we have found invaluable in business communication. If you're meeting with someone over lunch or a drink, there's often nothing more powerful than making a few bold lines on a cocktail napkin. There's long been a certain magic between the cocktail napkin and business plans. For example, CEO Herb Kelleher says that the business plan for Southwest Airlines started with a simple triangle drawn on the back of a cocktail napkin. My brother David once had his Stanford students write the "big ideas" for their class project on napkins, as practice for being future Silicon Valley entrepreneurs. In everyday office meetings too, I often find it easier to drive home a point by scratching out a few circles and arrows on a marker board. But of course, many of us haven't done any serious drawing since, say, the fifth grade. And when we try to draw people, we revert to grade school habits.

If you've got the stick figure blues, David's got a wonderful cure that he picked up from a Stanford professor. Whether you're a kindergartner or a senior executive, he teaches the same way. Make your usual circle for a head, then imagine you're drawing a droopy, bulbous leaf. Resist those crabby old stick arms. Let your pen sweep out like a four-pointed leaf, one for each arm and leg. I think you'll be surprised. In five minutes I've seen David teach the most art-resistant corporate types how to sketch a far more friendly, human-looking person.

HANDSHAKES

There's a good reason why I haven't talked that much about design. The classic image of the designer as the supreme architect of a product just isn't our style. Oh, we've got some of the best industrial designers in the world, and plenty of talented architects, but they always work in tandem with engineers and human-factors experts.

We work to bring the designer back to earth, and make him or her an integral part of a team that finds its inspiration in reality—how people actually interact with products, places, and ideas. This isn't about pretty colors, playful shapes, or style. To me, everybody can take the essence of good design and bring it to their business or products. Truly good design is about meeting people. It's about opening doors. Designers sometimes call these "affordances." I like to think of them as welcoming handshakes, visual clues that make you feel more comfortable.

We couldn't get through the day without good affordances. They're all around us. Think of red and green stoplights, the solid lines on the road that tell us to stay put, the broken ones that tell us we can pass, and the raised dots on the shoulder that just may save your life. A flat plate or a curved handle tells us whether to push or pull the door. A

weather vane tells us which way the wind is blowing, and, of course, Netscape's shooting stars let us know we're going somewhere.

Try boiling water in a teakettle that doesn't whistle, and you get some idea of what happens when affordances are lacking. The Paris Métro makes you push a button to open the door, and I've seen a few passengers whisked away to the next stop while waiting for them to open. Websites, of course, are in a class by themselves. Early ones were notoriously bad at the most basic communication. You were generally in the dark as to whether or not the site had finished loading, *when* you could click, *where*, and, *what* you might find. Nor is this a problem replaced by the gimmickry of bouncing images, streaming video, or rock tunes. Most sites don't have a clue about how to put visitors and customers in the driving seat.

Even the best companies can stray from their customers. We love the iMac, for example, but its first mouse wasn't up to Apple's standards. We've prototyped mice in just about every shape imaginable, including a small square that failed the human factors tests because it wasn't comfortable in an adult's hand and was too symmetrical for effortless orientation. We could have told Apple that a round mouse wouldn't work for the same reasons. The proof that Apple's mouse wasn't meeting people's needs (and lacked a critical affordance) can be seen in the thousands of people who have found that it only works with a third-party mouse holder—the Mouse Catcher, a plastic, oblong mouse-shaped device that snaps on top.

In fact, the entire computer peripheral business is proof that there's often something missing. From printers to expansion cards, from keyboards to every input device imaginable, manufacturers have raced to make the computing experience more complete. Handspring anticipated this wave with its open-architecture Visor, which assumed that even a tiny PDA can be expanded and improved upon. Then there are the thousands of people that turn their Palms into mini-laptops with a folding keyboard. Auto parts stores dedicate several aisles to products that fix things auto manufacturers either didn't bother to do or did

badly. Like seat belt adapters for kids or pulldown sun visors for people in the backseat.

Bad affordances surround us. They make us grouchy, they make us late. They give us the sensation that we're down in a drafty lobby, not knowing when to press the talk or listen button on the clunky old intercom. Think of how difficult word processors make it to perform the simplest task, like inserting a page break. Toys for grade schoolers that practically require an engineering background to assemble. Voice mail systems that befuddle us with too many choices. Childproof or tamperproof aspirin bottles that parents can barely crack open.

Find something lacking in a product, and you're likely staring at a tremendous opportunity. Ever notice how computerized registers at many retail businesses can actually *slow down* checkout? Couldn't leaf blowers be quieter, or trash bags be made to decompose? Why did e-mail lack spell checkers for so many years?

It took nearly eighty years after the invention of the automobile for manufacturers to start seriously coming to grips with the fact that Americans practically live in their cars and thus require a vast array of nooks and crannies (like the first adjustable drink cup holder) to store things. Backseat VCRs may be just the latest in this continuing evolution. As I write, Ford is experimenting with adding everything from washing machines to microwaves in its minivans.

Sometimes it's hard to tell whether something was really missing or whether you're just riding the latest craze. For example, at the turn of the millennium we did a translucent Visor for Handspring and a snappy translucent cable modem for 3Com. (Epson came out with a laser printer with a transparent cover that shows the paper's movement.) Some might argue that the wave of translucent products is a fad, but at its best it brings people closer to the workings of products that often seemed remote.

Look around your desk and you'll likely see plenty of evidence of homespun affordances: mouse pads with images, clips to hold paper to read as you type, funky screen savers to decorate your monitor. (Interestingly, the pads are not necessary except as they define your mouse

territory; similarly, screen savers may animate a dozing computer, but you no longer need them on many machines.)

Look at people or situations with unusual or specific needs, and you'll get a sense of just how great the opportunity is for strong affordances. Think of how Braille made reading possible for the blind. Or how shorthand made it possible for average typists to keep up with the spoken word.

In each case, innovators leaped a hurdle and reached out to people. Think of the remarkable advances in computers, laptops, and hand-held devices we've seen over the last twenty years. Mike Nuttall of IDEO made one of those seismic shifts in the early 80s when he decided the monitor didn't have to rest on your CPU or phone books anymore, and could instead pivot on a ball and socket mechanism. Later, he helped shift the architecture of mobile computers from a flat brick to the hinged clamshell, that is so familiar today. We hardly notice these little things, the way keys have gradually receded on laptops to give our wrists a place to rest, the way sounds have become second nature. But would the iBook be the same if it didn't cheerfully chime when it powered up or make those cool little vacuumlike noises when we cut and paste?

Good products or services are like generous hosts. They greet you at the door, offer you refreshments, and make you feel at home. I think it's a good thought to keep in mind, and it's not something you have to limit to the marketplace. Look around your office or consider your latest project. Is there a step you might take to make your ideas a little more approachable?

RITUALS REWARD

Just as Rome wasn't built in a day, rituals or traditions can sometimes create unusual challenges and opportunities. Rituals can cause us to hold on to the remnants of old processes that still resonate with

people. When Bayer started protecting its aspirin tablets with Toleraid microcoating back in the 1980s, the cotton balls packed in the bottle to keep tablets from breaking became superfluous. Yet it wasn't until the late 1990s that the cotton got left behind. Why? Customers wanted it. They said it reminded them of their childhood. There was also an enormous amount of myth wrapped up in that little ball of cotton—the idea that it kept the pills potent and soaked up damaging moisture. (Both notions are false, of course: Cotton actually draws in *extra* moisture from the outside air.) Pulling out the fluffy stuff had become part of the ritual of opening a bottle, something that told you it was new and safe. So strong was the nostalgia associated with the experience that only recently have Bayer and other manufacturers begun to ship aspirin bottles naked—without the cotton.

Indeed, how much change people will stomach is a moving target. When I was a boy, cake mixes tried to float the new, improved "just add water" versions. Homemakers revolted. It wasn't baking if they couldn't add an egg. So the cake mixes quickly retreated.

Try to replace a beloved ritual, and you'll find the barrier is extra high. Take the venerable wine cork. Everybody knows that fine wine needs a good cork, right? Only cheap wine has a screw top, right? So goes the popular mythology of wine and corks. The only trouble is that corks, while attractive and steeped in tradition, are a problematic way of maintaining the quality of fine wine. Over time, some corks fail. When the cork breaches, a tiny oxygen path opens and the wine starts to turn to vinegar. Sommeliers complain the wine is "corked."

Sean Corcorran at IDEO plumbed the depths of this phenomenon for a major wine maker. The brainstorms were wild. Sean bought, as he put it, "every bottle with a closure known to man," including obscure cosmetic bottles with plastic tops and quirky latches. A local wine store had a framed collage of the steps in the making of a champagne cork, so Sean managed to borrow it for the brainstorm.

The ideas ranged all over the map, including using so much of a nontraditional material, like plastic, that it seemed rich rather than cheap.

Or a combination cork and dispenser system. Oxygen spoils wine, but nitrogen preserves it. Sean sketched up a stopper with a onetime nitrogen charge you could dispense when you didn't consume the whole bottle.

After consulting with masters, Sean learned that metal actually makes the best seal. Utilitarian crown caps—like those used on beer bottles—would be far superior to corks in maintaining the quality of fine wine. The metal cap with a rubberized seal keeps out air far better than cork, and lasts longer. Was this a dead end? No, not necessarily. Sean hatched the clever idea of making beautifully stamped crown caps that would be elegant and collectible.

Of course, you probably won't find any of these ideas at your local wine shop anytime soon. No matter that wine corks sometimes seal wine poorly and that collectors must pay experts to periodically reseal rare wines. Over the years, we've developed an emotional, ritual attachment to every part of the process of opening a bottle of wine, from stripping off the lead wrapping (now aluminum) to yanking out, smelling, and sometimes keeping the cork. It's going to take considerable time and effort before wine corks are abandoned for bottle caps or an even better solution takes hold. Thousands of years of wine making have created one of the world's strongest rituals, a deeply ingrained mixture of myth, aesthetics, and habit that presents a formidable barrier to change.

We change our habits and expectations reluctantly, even when they seem to serve us badly. When I was in business school at Berkeley, I did a class project with a start-up company called Shoe Care. They said they wanted to be the McDonald's of shoe repair and bring professional management systems to a cottage industry. They pointed out that shoe repair shops were dirty and crammed from floor to ceiling with shoes that sometimes seemed impossible to find. So they took what seemed like a rational course: They broke that ritual with a pristine shop where the shoes were shipped off-site to be repaired, and then returned neatly wrapped. But when they opened their first shop

in San Francisco's Civic Center, people didn't know what it was. The venture failed because it neglected to build a bridge from the familiar to the new.

A similar fate almost befell Harley-Davidson, which nearly went bankrupt trying to compete with Japanese motorcycle makers in the late 1970s and 1980s. But new management decided to reach out to customers. They took an observation-based approach and rode to rallies to see how bikers were souping up their Harleys. They even hit the open road themselves for inspiration. They learned that Harley enthusiasts were angry about the motorcycle maker straying from its heavy chrome past. And so Harley-Davidson returned to its roots, playing up the outrageous size, fury, and ornamentation of the original Harleys (adding a little modern reliability). By the early 1990s Harley-Davidsons were outselling Hondas, and it had won over some of the most loyal customers in the world. (How many of *your* customers would tattoo your brand on their arms?) The company reinvented itself by celebrating its tradition and heritage.

Lego, of course, is that rarest of breeds, an old tradition that jumped into the high-tech age with robot-controlled constructions that rival the world's most automated manufacturing plants. Good rituals don't die, they just get reinvented. Take the elaborate seals and stamps that personalized documents decades or even centuries ago. Now people sport elaborate e-mail personas and append their electronic missives with titles, schedules, even witty sayings. Browsing the old library card catalog wasn't the fastest way to search, but it may have allowed more opportunity for serendipitous discovery than the computer. When you consider that rich tradition, I'm not surprised that some of the early, overly digital and logical methods of mining the Net aren't necessarily the most successful. Think of Yahoo, which uses humans to catalog the Web, and Ask Jeeves, which cleverly realized that people like asking questions in, God forbid, *sentences*.

Most baby boomers remember the transition from vinyl LP records to compact discs in the 1980s. CDs ushered in improved sound quality, greater durability, and simple storage. By trading tradition for inno-

vation, however, we did have to give up one small part of the music experience. Playing a record for the first time, listeners used to curl up with the oversized album covers, reading the liner notes and admiring the artwork. CD covers are just not the same.

In a similar way, we will go through a transition with hardcover books over the next decade. Everyone loves the feeling of curling up with a good book, and the prevailing opinion is that eBooks will never be the same. That's certainly true, but watch for eBooks to chip away at the ink-on-paper variety. Give us great prices, paperlike resolution, and new levels of interaction with authors, and most of us will switch over eventually.

Apple achieved that deeply textured emotional connection with its PowerBooks and iBooks, and I'd like to think we did it with the Apple mouse and the Palm V. It's what we strove for in our work on the leather-covered SoftBook electronic book. It doesn't happen often, but when it does, consider yourself lucky.

EVANGELISM WORKS

Putting energy and effort into a product, service, or idea can be the difference between a hit and a promising new product that doesn't quite catch fire. Where would the Macintosh be today without the enthusiasm and energy of Steve Jobs and his team of evangelists? The Macintosh wasn't sold so much as it was emoted, starting with that inspirational 1984 Big Brother advertisement that aired during the Super Bowl. Apple reached out to developers as well as consumers, knowing that without solid applications the machine would likely suffer the fate of its predecessor, the Lisa, which never caught on in the marketplace. Apple understood the importance of grassroots organizing and word of mouth. Macintosh user groups were embraced and made to feel part of a campaign, softening possible negative re-

sponses to the first model, which lacked, among other things, a hard disk.

We've met enough inventors over the years to realize the distance between a good idea and a successful product. Even if it's a true invention with little precedent, you'd better be prepared for some missionary selling.

A good idea isn't enough. Take the nuisance of coins piling up around the house. If you're like me, they collect like weeds. Banks don't really want them, and, besides, does anybody like putting change into those little paper wrappers? Entrepreneur Jens Molbak didn't, and he figured he wasn't alone. After earning his MBA at Stanford and interviewing grocery store shoppers in San Francisco, he realized he'd found a gap large enough to be filled by a business: The coins idle in American homes add up to several billion dollars. That's when Molbak came to us with the raw idea for a machine to exchange coins for cash in supermarkets. Nobody believed it could be a business.

We made five working machines that were installed in Safeway stores around the Bay Area. But this wasn't an impulse buy like candy by the checkout counter. The Coinstar only worked if you brought in your jars or sacks of coins. That made it at best a two-step process—see the machine, then bring in your stockpile of coins on the next trip. Slowly but surely, however, people started bringing in their piggy banks. Molbak got investors, a partner in manufacturing, and other stores to sign up.

Molbak had to do an enormous amount of evangelizing. The green-colored machine and its video-poker-like screen featured images of a jar of coins. "Get cash. It's as easy as 1-2-3," said a big display. But even those incentives weren't enough. Marketing people had to stand by the machine and tell customers, "Have you seen our new Coinstar machine?"

Ultimately, with lots of hard work and cheerleading, the Coinstar became a success. Today Coinstar has some seven thousand self-service coin machines in the United States, United Kingdom, and Canada and ranks as one of the nation's fastest-growing companies.

How did Molbak do it? He partnered with the leading grocery stores and created some bright incentives to encourage people to start the cycle—such as a Coins That Count program that funnels funds to non-profit organizations. In the spring of 2000 Coinstar hooked up with the U.S. Mint to promote the new Sacagawea gold dollar. Not a bad partnership for a company in the coin recycling business.

RUNNING WITH AL GORE

The importance of effort and clever persistence cannot be overstated. It's often not enough to work hard or follow what seems the logical course of action. Sometimes you've got to try a little ingenuity.

Several years ago, hoping to win an account to design some AT&T phones, David, Mike Nuttall, and I strapped on AT&T pagers. We set up an 800 number for AT&T executives to call, and modified one of their phones so that it had just one big button—a direct line to IDEO. We were willing to answer questions from them day or night. They got the message that we were serious, and appreciated our playful approach. We won the account.

Rich Silverstein of the celebrated advertising firm of Goodby & Silverstein may take the cake for the most audacious effort to win over a potential client when he pitched Isuzu Motors. Silverstein printed up flyers and had them placed in the windshields of Isuzus all over San Francisco, offering to pay fifty dollars to anyone who parked his or her car within a two-block radius of the advertising office at the foot of Telegraph Hill. Who could resist? The visiting Isuzu executives were amazed to enter a neighborhood that looked like a giant Isuzu dealership, and Silverstein got the account.

Persistence pays off. You won't always succeed, but if you keep trying, eventually you're going to get through. Some years back, a few IDEOers caught wind that Al Gore was coming to Silicon Valley. Dave

Blakely and another IDEO engineer, both serious runners, figured they'd try to join him on his morning jog and invite him over to IDEO. I can imagine a hundred reasons why Blakely's scheme wouldn't fly, one of them being Secret Service agents. But Blakely is used to making things happen.

"You don't think you're going running with the vice president, do you?" a Secret Service agent told Blakely when he arrived at the San Jose Hilton shortly after dawn. No one was allowed within eighty feet of the vice president. But Blakely trotted after Gore and his entourage and a mile later was running and chatting with him. Gore had just been on the Letterman show, so Blakely pitched him ten reasons, a la Letterman, why he should visit IDEO. Gore, as it turned out, was too busy at the time to schedule IDEO. But to me that's not the point. If your staff is regularly dreaming up creative ways to advance your company, you're sure to break through eventually.

When somebody champions an idea, it's amazing what a difference that person can make. I couldn't begin to recount the times we won a job or successfully finished one because someone decided to go to the mat for the project. No matter how cool your company may be, we all get tired and sometimes feel we're lost in the crowd. But match up a little bit of creativity with some persistence, and you might just be the person who changes the whole darn company. A simple idea could transform the way you're perceived. Like getting the American Dental Association to sanction your toothpaste or Leonardo DiCaprio to wear your sunglasses.

You could have that idea.

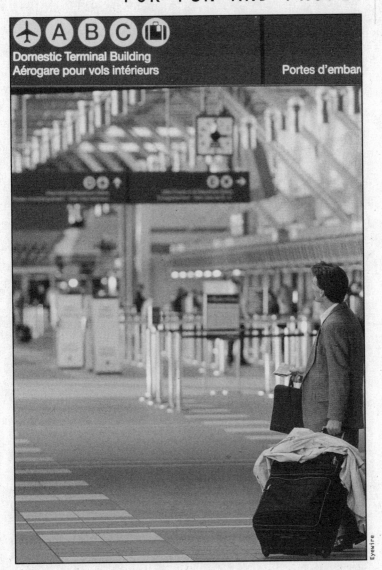

Domestic Terminal Building
Aérogare pour vols intérieurs

Portes d'embar

Eyewire

Innovation is a very goal-oriented process, and hot groups under tight deadlines tend to focus closely on end results. As you step through the innovation process, try thinking *verbs,* not nouns. What does that mean? It means not focusing too much on the object or artifact: the new product, the big report, the latest ad campaign, the remodeled store. Everybody's in the business of creating experiences, so focus on the verbs, the actions. The goal is not a more beautiful store. It's a better shopping experience. And creating more value for your brand.

Look around and you'll see innovative companies and powerful brands linked with the word "experience." The Lexus experience, the Nordstrom experience, the Webvan experience. What's more, if you have been a customer of those companies, you have a clear idea of what is meant in each case. Lexus makes great cars, but the "experience" is more about pampering and courtesy and not just a smooth, easy ride. Nordstrom sells clothes, but people don't talk about their Nordstrom garments, they talk about the piano music, the friendly, attentive help, and going the extra mile. Webvan sells the ultimate commodity—groceries—but what it's really known for is convenience and timeliness.

How about your company? Are you an innovator in delivering experiences—either to your staff or to your customers? As you go through the innovation process, there are plenty of opportunities to seek out the verbs.

You can turn almost any product or service into a better-designed experience. When Amtrak asked us to help design a train interior for its Acela service between Washington and Boston, we came back with ten unique steps in a journey, everything from planning a trip to arriving at your destination. The key was recognizing the importance of moving beyond the physical train. Helping Amtrak redesign the inside of the train car was just part of the Amtrak experience.

GOING OFF-BROADWAY

In *The Experience Economy* Joseph Pine and James H. Gilmore argue that many businesses and services are coming to resemble the experience of a theme park. It's a concept, they believe, that is spreading rapidly throughout all kinds of industries. The day isn't far off, they say, that we'll actually find ourselves paying for the privilege of going to premier malls.

In explaining the evolution toward experience economies, they use the metaphor of a child's birthday cake. In a commodity-oriented economy, Mom bought flour, sugar, and eggs, adding all the labor herself to bake and ice a cake. The next evolutionary step—a product economy—has people buying Betty Crocker cake mix and a can of icing. In a service economy, the cake is made in a bakery, and they'll even put your child's name on it. But in an experience economy, you take Mikey and his friends to the local Chuck E. Cheese or Discovery Zone, where they are provided with active play, pizza, drinks, and—oh, by the way—a birthday cake at the end.

Think about experiences that you might stage for your business. Sales teams, with their focus on results and rewards, have always done this on a grand scale. Many companies have some version of a President's Club. When I was at General Electric, the reward event was called the Mark Makers. I was amazed at the over-the-top experiences they designed for their salespeople. First, they'd fly three hundred of us and our significant others to Hawaii. The theme parties were pretty impressive. One year they converted the entire grounds of a lavish hotel into a MASH unit and dressed us up in fatigues. Another year they flew in Monty Hall of *Let's Make a Deal* to give away trips to the Bahamas and cars to the top producers. Within days of returning home, professional-quality photos of you and your spouse showed up in the mail. People talked about the event for months.

OK, so how do you create a designed experience for a new product or service? Here's a hypothetical example of the kinds of steps we often go through.

Take an experience you're familiar with. Say, air travel. If you've flown lately, how would you describe the experience? I'll bet you noticed something that could be improved. When we design an experience, we add up all those small incremental improvements into a better complete experience.

One way to begin is to break the product or service down into its component elements. The DNA of the experience, as it were.

What are the main steps in taking a flight? In simple terms, you plan the trip, deal with the airport, board the plane, fly, and arrive at your destination. Unfortunately, it's hard to get insights from those general steps. You have to go to a much finer level of detail. For instance, here are some of the steps involved in just one part of the journey—the airport experience:

1. Arriving at the airport by car, taxi, shuttle, bus, or train
2. Finding the right terminal
3. Checking in and checking luggage at the counter
4. Confirming or getting your seat assignment and departure gate
5. Clearing security
6. Locating and reaching your gate
7. Checking the boarding status and updated flight schedule
8. Making upgrade or standby transactions
9. Deciphering preboarding announcements
10. Walking down the jetway

What's the dream scenario? James, your limousine driver, wheels you through a special security gate out onto the tarmac where your private jet is waiting with its engines running. While James loads your luggage, you step onto the plane and the doors close behind.

For a few dozen corporate CEOs and celebrities, life is like that, but it probably doesn't match your recent flying experience.

So what's the more typical scenario? You have probably not been hit by all of these disasters on the same trip, but lots of this will be familiar:

You leave plenty of time to get to the airport, but get bogged down in traffic, misread the exit sign, then can't find your terminal and have to circle the entire airport. Your problems have only just begun. A ballpark-sized line snakes nearly to the terminal entrance door. Though you're not sure what's the right line, out of desperation you jump in, only to learn ten minutes later you're in the wrong line.

When you finally reach the counter, the agent either disappears into the back room or ignores you and spends several irritating minutes helping the confused agent beside him with someone else's botched ticket. Next, you wait in line at security for the chance to hoist your carry-on bag onto the conveyor belt, empty your pockets, hold up your arms, and be treated like a criminal.

After some minor confusion with the airport signage, you find your gate—which, of course, is at the end of the most distant spoke of the terminal. You hear on the public address system that a flight is in the final stages of boarding, but you can't tell if it's yours. Although your flight seems ready to leave, there's no one to ask whether your section has boarded. In some versions of this scenario, you actually make your flight. But as you're kept waiting until the last possible moment for an upgrade, you worry (correctly) that the overhead bins are filling up.

Guess what? You lose. No upgrade, and now you've got to check your laptop.

Let's face it. A fair amount of the airline experience isn't what it should be. I believe that customers will reward the airlines that strive to systematically improve this process. Here are some possible ways IDEO might redesign parts of this experience:

Your color-coded ticket features an icon that corresponds with the departure terminal, and the outdoor signage itself is colorful and memorable. Inside, customer service agents and well-placed signs make it foolproof to find the right line. While waiting in line you check in at a rental-car-style mobile kiosk so that when you hit the counter, it only takes seconds.

In security, a floor-level conveyor belt effortlessly sends your bag through the X-ray machine without you having to hoist it up.

A smart ticket envelope (you got at the ticket counter) with a color-coded map directs you to your gate. There, a striking revenue-generating display (funded by advertising) tells you what's up with your flight. It's on time, and rows 30 to 50 are currently boarding. You can see the number of seats available for upgrade and how many people you're competing with for a seat. Enter your ticket in a mag stripe, and it shoots out your upgrade number for the waiting list.

OK, it's time to board. A large monitor over the gate door tells you instantly which rows are boarding and how many minutes remain before the door shuts. A flashing icon on the display unequivocally announces when late arrivals forfeit their reservations and seats become available.

This is all hypothetical, of course, and if we were working for a client, this would just be a start. Still, I'd argue we've already mapped out some general ideas about how to design not only a smoother, more hassle-free experience but one in which the traveler has more control.

You probably don't work for an airline, but you've seen how you can begin to ask the basic question. How can I redesign my operation, install the latest technology, or retrain my staff to deliver a better experience—without, of course, breaking the bank? Start by following your customer journey, breaking it down into its component elements, and asking yourself how you can deliver a better experience.

It's not all about intensely complex services or products.

We've found that the smallest devices or services can inspire well-designed, worthwhile experiences. Years ago, for example, Bell & Howell asked us to look at the inventory control systems it sold to auto parts mechanics and auto service centers. We observed that the typical sales process involved the clerk asking the customer if he'd pulled up the right part. Trouble was, the customer had to lean over the counter to glimpse the display, a cumbersome process at best. That awkwardness led us to design a display that could pivot 360 degrees, so the clerk

could swing the part description and diagram toward the customer (a capability we evolved a decade later for NEC laptops). It's a small innovation. But it significantly changed the way customers felt about the sales process.

Innovation sometimes starts with a small improvement. But a better-designed experience often comes about when you can transform a niche product to something broader that resonates with your customers.

Merloni, one of the largest appliance vendors in Europe, asked us to design a tiny controller for the home that would help customers manage their power usage and shut off appliances before fuses could blow. Electricity is extremely expensive in Europe, and if Merloni could help customers use its machines more efficiently, and automatically notify its service department when something needed attending to, the company was bound to strengthen its brand.

Early in the project, however, an IDEO team led by Douglas Dayton realized that the device Merloni was considering would require so little interaction that customers would forget about it. Nor would they likely see much added value. We advocated a new approach. Why not use this minor product as part of a home communications hub, providing calendar and time management, shopping lists, food ordering, weather channels, and radio? We hooked up a device we designed to the Internet and showed how Merloni could recruit content providers to offer its customers a variety of services. The resulting Home Smart Monitor still accomplished the original power leveraging and service functions. But the breakthrough was adopting active, vital usage scenarios that would be attractive to families. That didn't just give the device a raison d'être; it also lit the path for new business models for Merloni—that of service provider for some 7 million European customers.

A healthy portion of our work involves exploring new uses for existing technology or services. This is something nearly any company can do—and it can reap rich rewards. More and more of our work seems to

reinforce the notion that no matter what the product or service might be, ultimately it's about some sort of human experience. When we worked with Medtronic on new systems for implantable defibrillators several IDEOers were asked to gain some firsthand insight into the cardiac patient's experience. I wore a pager for an entire weekend, and each time it buzzed out of the blue it signaled that I'd just gone into fibrillation, and received a powerful shock to restore normal heart function. It made me see how easily you could be hurt if you fell to the ground, and how many hard objects surround us. One time my beeper went off when I was alone in my garage operating a table saw, a scary situation for cardiac arrest and the jolt of defibrillation that it would trigger. Other team members carried the beepers for a weekend as well, and the exercise gave them new insights to help them make better products and services.

Consumer technology leaders recognize that on some level they're no longer in the gadget business. Recently, we worked with Nokia and several European universities to explore how mobile technologies could be used to *increase* rather than reduce face-to-face interpersonal communication. In some markets, coming up with compelling reasons for people to use your product or service is a question of *survival*. If you don't, you risk becoming as obsolete as the typewriter.

MAKE EXPERIENCES ENTERTAINING

While the experiential approach I'm talking about is fairly new to Internet devices, cell phones, and computers, it's a tried-and-true concept in the retail and entertainment industries. Major retailers were among the first to recognize that they were selling themed entertainment as much as the merchandise on their shelves. I think there's a lot to learn from their successes and failures.

Roberto Pisano

REI's sixty-five foot climbing structure offers an
experience on-line retailers can only envy.

NikeTown, Warner Brothers Studio stores, and Disney Stores sell themed Experiences with a capital "E." Each NikeTown celebrates sports, with roughly half the retail space devoted to taped sporting events, memorabilia, and displays. Disney Stores calls its employees "cast members" and makes them part of the studio-lot experience, while Warner Brothers shoots for movie-set ambience. At the Levi's store in San Francisco, a futuristic machine scans your body with laser beams to measure for custom-fit jeans; more exhibitionist shoppers can soak in a public tub to shrink their jeans to fit. The REI flagship store in downtown Seattle features a 65-foot-high climbing "mountain," a 470-foot-long hiking trail, thirty pitched tents, a Rain Room to check whether the gear is truly waterproof—even a dirty pool for campers to test water purifiers. Aisles are made to resemble hiking trails, and door handles are made of ice axes. It's a dazzling mix of entertainment and education. Moreover, the staff members are themselves mountain climbers, cyclists, and kayakers, able to dispense expert advice to customers. And with 1.5 million visitors in its first year, the store earned its own designation as a National Park Service information station.

It's easy to understand why mass-market retailers would create such big stages for their products. But what about the hundreds of selling experiences that take place in corporate America every day?

A few years back, Steelcase asked us to put our understanding of designed experiences to work on a big stage—the heart of Manhattan. Steelcase, like the nation's other major office furniture makers, had long had a huge showroom at the International Design Center in New York. But when the spot lost its cachet, the company started looking for a new showroom and focus.

When Steelcase executives started working with our environments group, the success of NikeTown-style experiences was fresh in their minds. Steelcase was looking for a way to improve sales, of course, but it also wanted to demonstrate its expertise and knowledge of workplace practices. Contract furniture companies have generally taken a tradi-

tional approach to selling their wares. They usually had a display show-room out front and a back-of-the-house office where the staff worked. Buyers would be shown the furniture, led into a conference room, handed a refreshment, weighed down with mounds of literature, and sent on their way.

After intensive observations of Steelcase's customers, the team came back with what we called an "experience map," breaking down the range of buyers visiting the showroom and the potential points of contact. Right off the bat, we identified one of the great failings of most showrooms and corporate offices. Nobody likes to be kept wait-ing in the downstairs lobby only to be sent upstairs to wait once again. At Steelcase's New York Work/Life Center, waiting was eliminated. Buyers were immediately escorted to a comfortable seat in the elegant café and served refreshments, fruit, and even lunch, while enjoying one of the best views in New York, overlooking Central Park. Visitors could then take a trip throughout the space. The team created starter kits for buyers, complete with background documents and postcards that captured the flavor of the furniture. Even the elevator told a story. On each floor was a different element of a vertical time line dating to the 1920s, reinforcing Steelcase's long record of success. Polaroid cameras were provided so that you could photograph a particular piece of furniture or installation to create your own document.

While not a theme park per se, there were ample opportunities for interaction. "Sandboxes" invited buyers to try a chair or desk on for size, with fabrics and materials arranged for easy viewing. "Product pods" showed sample work settings, offering the interplay of infra-structure, technology, and ergonomics. Laserdisc case studies, mean-while, showed real office scenarios. But the biggest change may be how the Work/Life Center brings the back-of-the-house functions front and center. The workplaces of the salespeople and support staff became an integral part of a visitor's journey, making it possible for buyers to see the furniture in use. And after the tour, the buyer could return to the café and enjoy a snack with the salesperson.

In bottom-line terms the space has worked—sales are up dramati-

cally, and the average buyer spends more time in the Work/Life Center than in a traditional showroom. Equally exciting is that it has created a more vital workplace—the Steelcase employees in the building also enjoy the café with a view and feel invigorated by being an integral part of the showroom.

LEARNING FROM VEGAS

You can't design better experiences by staying in the office. You've got to walk designed experiences with your own two feet and see them with your own eyes. There may be no better place to observe how effective a designed experience can be than in modern-day Las Vegas. If you haven't been recently, I highly recommend a trip—for business, of course. I'm dead serious. No matter what your field or industry, whether you sell retail or business-to-business, you can learn a lot from Las Vegas.

The original Vegas casinos made you trudge by hundreds of slot machines and gaming tables to get to your hotel room. It was not happenstance that there was no place for the weary to sit except at a blackjack table or a slot machine. Vegas abolished the outdoors and day-to-day reality. Without windows or clocks, casinos made it almost seem natural for time, and your money, to fly. Search for a water fountain to dilute the alcohol-induced haze, and you'd come up empty. Even the underpriced rooms and budget buffets were all part of the equation, a sucker's bait and switch.

The old Vegas crumbled, of course. Treating people more like cattle than humans usually fails in the end. What Vegas did was to reinvent itself, and today it's the ultimate international virtual reality. In one afternoon you can dine in the Eiffel Tower, float down the canals of Venice, surf six-footers in the South Pacific, and walk the streets of 1930s Manhattan. The old objectives of keeping people awake

(and gambling) are now blended with opportunities for shopping and fine dining. Caesar's Forum pioneered the widely copied faux ceiling that mimics the rise and fall of the sun over and over again long into the night, and much like natural light, keeps you going into the wee hours.

The latest creations—such as the Venetian, Paris, or Bellagio—do a good enough job of approximating the experience of visiting European cities that many Americans who don't know or can't afford the real thing consider it a reasonable facsimile. Steve Wynn's Bellagio, modeled after the town on Italy's Lake Como, may be the largest, most elaborately designed luxury environment in the world. Seven celebrity chefs have lent their names (and sometimes their talent) to the Bellagio dining experience. The elegant boutiques feature only the ultra-premium brands, and the art museum includes portraits by Cézanne, van Gogh, and Picasso.

The Vegas experience is still about "going over the top," but it's no longer only about outlandish neon and flashy dinner shows. Real Italian cypress trees and Aleppo pines line the Bellagio's pool. The French may turn their noses up at "Paris," but it's not hard to imagine you are in France when you are there. Most of the casino's four thousand employees speak a little French, and if the baguettes taste authentic, it's because the oven was shipped from the homeland. Even the paint used on the half-scale replica of the Eiffel Tower is identical to that used on the original.

MAKE THE HUMAN CONNECTION

So what can Vegas teach us about other business "experiences"? There's nearly always more opportunity to create excitement than you think, whatever it is that you make or sell. Central Market, the Austin grocery store, for example, has turned shopping into a trip to a "foodie"

amusement park. Central Market's 63,000-square-foot space offers a Vegas-proportioned experience. Customers will find two hundred varieties of olive oil, five hundred cheeses, and thirty kinds of sausage; the butcher counter alone is half the length of a basketball court. Breaking from the monotony of standard aisles, Central Market has boldly created a yellow brick road-like path that winds through the store, creating more shelf space and giving customers a grander view of its bountiful produce, diverse foods, and lavish displays. Every kid gets a balloon, and official "foodies" offer free samples and are given carte blanche to open up any package for sale and offer a taste. Cooking students "Shop the Market," gathering exotic goodies and whipping up gourmet meals on-site. And when you order at the café, you receive a pager that brightens when your meal is ready.

If you're thinking food has nothing to do with your business, you're probably wrong. When was the last time you had lunch catered into a conference room for a client? Remember how their eyes rolled when you offered cellophane-wrapped sandwiches that looked and tasted like they came out of a vending machine? When we order lunches for our staff and clients for brainstorms and other meetings, we try to make it fun and delicious. It can get you off on the right foot with a client.

Bon Appetit, our favorite caterer, does a great job of changing the experience of a box lunch by redesigning the box. For one special event, they packed a "Nordstrom picnic," with each beautifully presented meal in its own silver-and-gold-colored Nordstrom's gift box. Another time our sushi lunch arrived in a flower box suitable for long-stem roses, with a long plastic window to show off the food.

We all crave good experiences. It's one reason "experience gifts" are on the rise, with people giving everything from wine-tasting classes to tennis lessons or a weekend at the spa. Companies have gotten better at giving experience gifts to employees as well, from the sales conference in Hawaii to the off-site at a resort or a team trip to a sports event.

By turning your product or service into an experience, you just

might find a market you didn't know existed. The Hank Haney Golf complexes in the Dallas area used to see revenues slump every December. Then they started advertising their gift certificates for golf lessons and driving range experiences on local billboards. Suddenly, that new business climbed to over 50 percent of their holiday revenues and put December in the black. Customers have the chance to give an *experience* gift, and the golf courses get to remove a dip from their seasonal revenue curve.

Designing new experiences is usually about figuring out a way to connect with people. There was a time, for example, when all the travel industry offered was air travel, hotels, and rental cars. Now travel agencies offer adventure travel packages, eco-travel, art travel, and literary travel. If you've got a special interest, there's undoubtedly a travel company that offers a travel "experience" for you in some corner of the world. And the adventuresome companies guiding people through these experiences are reaping great rewards. Though tourism plods along with less than 5 percent annual growth, nature and adventure travel is up by 30 percent a year. Who is responding to these new experiences? Millions of people seeking authentic or meaningful experiences. Consider how radically eco-tourism has redesigned the scope of travel. Other than joining the Peace Corps, you probably can't get a much more meaningful travel experience than eco-tourism, vacations designed to help sustain the environment and local inhabitants.

By adding a little spice to what you produce or sell, you may have people lining up as if they were tourists going whale watching. The Pike Place Fish Market in Seattle is a good example. How can one breathe new life into the traditional experience of buying fish? Well, Pike Place Market did it in spades. You point to the fish you want, and then, as the fishmongers start hollering back and forth, one of them winds up and pitches your shining ten-pound salmon twenty feet or so to another guy holding a makeshift catcher's mitt fashioned out of wads of newspaper. They neatly wrap it in an ice pack for your drive or flight home. It works! It's a dazzling performance—there's nothing like

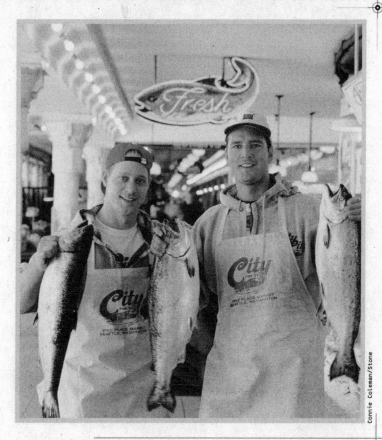

Connie Coleman/Stone

Even buying fish can be
turned into an entertaining event.

flying fish to draw a crowd. On the surface, nothing could be more
mundane than selling fish, and yet Pike Place Market has designed it
into an experience so compelling that Charthouse Learning actually
wrote a book—and best-selling video—about it.

Think about the process involved in making or delivering your com-
pany's product or service. There's always an element of the experience
you can bring to life. People have been making tortillas in Mexican

restaurants for ages. But the Fresh Mex restaurant chain Chevys realized that if they brought the tortilla wheel—dubbed "El Machino"—out of the kitchen, customers would be entertained by watching them being made, all the while reinforcing the fact that they are getting fresh tortillas.

Can you put a new spin on the "tortillas" of your office? How could you turn your *invoice* into a fun experience?

At IDEO, our hands-on front lobby displays turn waiting into an entertaining experience.

How important is a good experience? Just look at Kinko's—they gave telecommuters, freelancers, and other mobile businesspeople the resources of a big corporation right in the local mall. All sorts of business functions previously available separately—custom printing, computer services, even meeting rooms and videoconferencing—were suddenly brought together under one roof. The timing was impeccable, coinciding with the estimated 40 million Americans who now work out of their homes, the rise of desktop publishing, and the Internet. To the "free agent nation" of entrepreneurs and telecommuters, Kinko's is the satellite office, offering the latest technology and services. Proof of the success of this designed experience is in the pudding. Customers now spend hours computing, meeting informally, or surfing the Net at Kinko's stores—and, of course, making copies too.

TELL A STORY

The next time you travel, try slipping off the beaten path. I believe in learning from both ends of the spectrum of the designed experience—from the Vegas and Disney side of the equation as well as the local dive. There's plenty to glean from one-of-a-kind spots. Often success comes from taking a decidedly low-key approach. We all know and love our local breakfast dive with its big portions, lots of noise, and take-it-

or-leave-it waitresses. We wait outside in the outrageous line because it's a good place to read the paper and soak in the atmosphere. Part of the appeal is that we feel such spots are "authentic"; we'd never queue up so long to dine at a chain restaurant.

The success of the Martha Stewart empire is largely about her savvy understanding of our desire for authentic experiences. Whether she is offering an illusion or not isn't the point. Stewart's hard-to-master crafts and complex recipes are about the nostalgic appeal of a different era with its own values and sense of time.

Well-designed experiences strike a nerve with the public. Take the story of the popular Restoration Hardware chain. It began, as so many good companies do, with someone trying to solve what seemed like a little problem. Founder and CEO Stephen Gordon was trying to open a bed-and-breakfast. But he soon found it a challenge to locate the period hardware and authentic lighting fixtures he needed to restore his Victorian inn. As he gradually cultivated national sources, friends and neighbors began asking how they could enjoy his new pipeline. Gordon wisely scratched the B&B and in its place hatched Restoration Hardware.

Just as Vegas has created a discount trip to pseudo-Europe, Restoration Hardware offers the serendipity of searching through an estate sale without the risk. Entering a typical store's "great" room, you wander from section to section, finding Art Deco–inspired chairs, Victorian glass doorknobs, Le Zig-Zag corkscrews, and even vintage toys. Except that everything is new. The store's natural lighting, architectural detail, and fine cabinetry underscore its focus on quality and craftsmanship. Surprise is key. Gordon loves juxtaposing ten-dollar novelty items with two-thousand-dollar leather chairs; return the following week and you'll likely find a new eclectic treasure. Most of all, Gordon tells personal stories about things. He approves every one of his store's more than 4,500 items, introducing them with placards carrying his tales of the worthiness of the offerings, often laced with his personal memories.

Restoration Hardware may be as much about a chance to buy a

piece of the traditional good life in the Internet age as it is about well-crafted hardware and furnishings. But doesn't Starbucks offer the same momentary "good life" refuge with its careful balance of good coffee, music, and wood-grained ambience? All around us the best stores, shops, and restaurants are drawing us in with "stories," whether romance, nostalgia, or mystery.

If you're scratching your head wondering what this has to do with your business, think again. Does your company host events, introduce products or services, and meet clients in public places? Most companies tell visual stories on the flat, one-dimensional level of a white paper or press release. If you've ever been to a giant trade show in Vegas, you know that certain booths and parties are packed, and others are painfully empty. It helps to have something worthwhile to sell, but selling is also about designing an experience for potential clients and customers.

FIX IT

We're fortunate to have worked with some of the pioneers in high-tech, legends like Apple and Dell Computer.

Along the way, we've tried to glean a little bit of what these companies do best. Just a few years ago it was incredibly frustrating to piece together a new computer. Even though I travel with a Dell machine, I still believe nobody can match Apple's out-of-the-box experience. Well-designed screen icons and manuals that are beautiful and almost fun to read make it easier to put together an Apple than the average children's toy. When the iMac came out, I realized the Internet-friendly machine was ideal for my computerphobic parents and my new-to-computing toddlers, so I bought two machines almost immediately.

All sorts of companies could learn from Apple's superior out-of-the-box experience. You have to wonder about all the "easy-to-use" software programs that spawn stores full of 500-page, $35 books to help you use them.

Dell successfully tackled two other key parts of the buying experience by pioneering the introduction of efficient, direct sales and by making the configuration of memory and peripherals nearly as easy as asking the deli counter to pile on the mustard and hold the mayo. A direct, reliable customization experience became Dell's strongest selling point, a competitive advantage that helped to make it an industry powerhouse.

Lexus, of course, has fixed all sorts of things that are broken, but being forgetful, I'm most grateful for one of its simplest, redesigned experiences. The experience of locking your key in your car. Leave the key in the ignition of a Lexus and lock the door, and the doors will shut for an instant and then release. If you somehow miss that safety feature, they also give you a credit-card-sized key to keep in your purse or wallet. Lexus knows you're not keen on the experience of locking yourself out.

When we redesigned the shopping cart for *Nightline*, we began our research by investigating shoppers' experiences with the traditional cart, everything from how people shop to how they navigate the aisles. By fostering "hub shopping"—carts with removable baskets—we tried to change the way shoppers experience the store. Our objective was to fix a broken experience that has barely changed in half a century.

SAVE A LIFE, TAKE A JOURNEY

In some cases, fixing broken experiences can mean the difference between life or death. That's why we designed a defibrillator for Heart-

stream to be as easy to use as a fire extinguisher. It's portable, features childlike 1-2-3 screen instructions, and has physical and visual clues to make its use nearly automatic. Since its introduction in 1999 dozens of lives have been saved with the device.

Designing new experiences is about striking a balance. Toss in too many new features, and it's like a movie that's too long or has too many characters. Medtronic's first redesign of its temporary pace-maker boasted impressive advanced features. But to run it properly, you needed a one-hour training seminar and a half-inch-thick manual. Each button had multiple functions in order to pack in lots of options.

Medtronic quickly realized the new pacemaker wasn't selling as well as it might, and asked us to rethink the process. We weren't content with the numerous interviews with doctors that Medtronic had conducted. We observed doctors and nurses at work. We noticed that, 80 percent of the time, the medical staff was using just a few basic functions. So we came up with three dial knobs with strong visual graphics showing default settings. Probably our biggest contribution was a layered menu approach. Cardiologists could still use deeper layers of the device, but the typical caregiver was no longer confused by too many complex capabilities. The best products and services aspire to the classic design principle "Make simple things simple and complex things possible." Sometimes designing a winning experience is about reining in your wish list and resisting the temptation to do too much.

RETHINKING SERVICES

Talk about broken experiences. Think of the typical hospital room, not the place most people are eager to check into. Precisely why Wellness

LLC, a Nashville-based company that designs and installs modular hospital patient rooms, is studying ways to make its rooms promote healing, adapt to the latest technological innovations, and increase staff efficiency.

Wellness's futuristic patient rooms are designed to be easier to clean and maintain and have been described as being more "holistic," featuring rounded surfaces and calming colors. The company even claims that the experience of installing the rooms takes less than half the time of conventional construction.

Established industries are ripe for a little redesigned experience. Nothing is too big or too small to benefit from a fresh perspective. Imagine how many service companies could use a new lease on life. The hotel industry is a prime example of how broken an industry can be. It also illustrates how quickly leaders can churn out newly designed experiences. Years ago, for example, the Mirage casino developed an airportlike computer-controlled hotel baggage system (aided by teams of bellboys) that can ferry your bag to your room almost as fast as you can get there yourself (while saving you the time and embarrassment of lugging your baggage through the casino). Some Las Vegas hotels, undoubtedly trying to increase the time you have to recoup your losses, have made it possible to check in with your departing flight before you leave the casino. Ten years ago it might have sounded far-fetched to think you could pay for your hotel bill through the in-room TV. But how far off can the day be when most hotels will give you the option of checking in electronically with a swipe of your credit card at a lobby kiosk and speeding straight to your guest room?

Look at how successful hotels have gone cyber, beefing up rooms and business centers with computers, Internet service, and e-mail. Luxury hotels, of course, live or die on how well they cater their services to the needs and desires of their guests. Consider these whimsical offerings by the Park Hyatt Los Angeles: the "movie turndown," with popcorn, candy, soda, and, of course, choice of movies; the "pic-

nic in the park," with the usual goodies in a basket; and the "bath turn-down," complete with bath salts, rubber ducky, and loofah.

LITTLE EXPERIENCES COUNT

What can we learn from studying *badly* designed experiences? Look around your own workplace or business for routines that people hate, established ways of doing business that burden your company and/or clients and customers. Little improvements matter. Do people enjoy returning something on a warranty? I sincerely doubt it. That's why when a customer returns a backpack for repair to JanSport, the company mails out a wacky response. "Hi! Warranty Service Camp is really cool. They say they're sending me home soon," reads a typical post-card. "Gotta run . . . we're doing zipper races today!" Silly as these cards may sound, JanSport reports that instead of getting mostly angry customer letters, nearly all the correspondence now consists of cheery thank-you notes.

IDEO may lack a warranty service, but like a lot of companies, we get our share of résumés from folks looking for employment. The old way of dealing with the tide of résumés made little sense. The count-less people we didn't have a place for or who didn't fit our requirements were mailed dreary attorney-approved rejection letters about a month after they applied. But the more we thought about this, the more we realized that we were typecasting ourselves as the bearer of bad news, not exactly the best course to create good buzz (or karma, for that mat-ter). Why not change the model and our image? We adopted a policy of promptly mailing colorful postcards, letting people know we'd got-ten their résumés and appreciated their thinking of IDEO. The tongue-in-cheek card begins with "Thanks, we received your:" and then gives a list of items with checks marked off next to what we'd been sent—letter, résumé, portfolio, or even their parking ticket.

But it was serious too, and it set expectations for job seekers. "We review all employment inquiries very closely. If there's an immediate match between your talents and our needs, we will contact you directly within four weeks. (It can take some time for a résumé to make the rounds of all the people who need to see it.)"

We'd short-circuited a tired old routine. Not only was a rejection letter unnecessary, but we also included some wording to let people down easily. "If there is no fit at this time, we will keep your résumé on file—honestly, we will. We have several valued employees who applied to work here more than once! Please let us know any significant changes in your experience or qualifications."

To me, this is a designed experience with a "small d," the sort that anybody (with an open mind) can bring to their group, a division, or company.

"Small d" designed experiences don't have to be about your primary products or services. Take company security. Few would disagree that it's notoriously bad and uninspired at most firms. You sign a book and stick on a badge. I've gotten interesting and occasionally valuable information by seeing who signed in today and who else my client is considering. It's not hard to imagine a scenario in which someone could learn a lot simply by flipping through that sign-in book. In the process of protecting their assets, companies sometimes compromise their secrets.

There's also a glaring contrast between the typical, nicely laminated employee badges that most companies issue, and the cheap paper ones that visitors get—even executives with whom you're hoping to do millions of dollars of business.

None of this struck us as secure or smart business. So we brainstormed a new model, to treat visitors as travelers that need a passport or tourist visa. And so "the IDEO Tour" badge, featuring a tourist wearing a Hawaiian shirt and lei with the Statue of Liberty, Big Ben, and the Golden Gate Bridge in the background, was born. We snapped tiny Polaroids of visitors and glued the head shots inside this fun tourist

frame to personalize the badge. On the back we printed a colorful map of the IDEO offices. Visitors wore it like a necklace with a nice metal clip.

OK, I admit it—sometimes we sound like kids that haven't grown up, making up games on a summer afternoon. It's easy to stop yourself and feel a tinge of guilt as you wonder whether you should be having so much fun at work. But try resisting the urge to stop. Who knows? You might just find that spirit of fun will push you toward designing a great new experience.

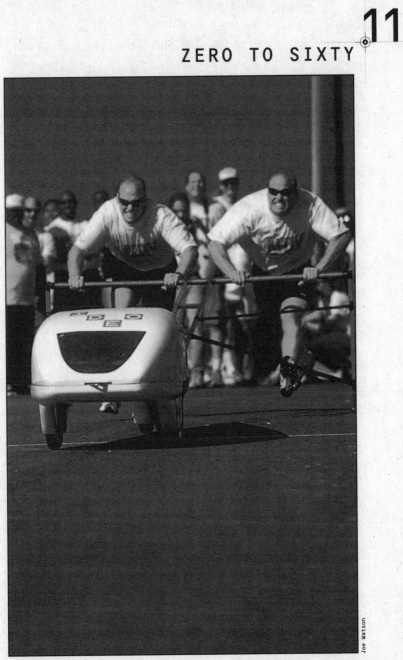

Joe Watson

Speed counts. It gets products to market in time for the holiday shopping season. It leapfrogs established players before they can get off the ground. It implants your brand in consumers' hearts and minds before your competitors have left the starting gate.

You need great teams to make innovation second nature in your company. And great space. But even if you've got the right people and place, innovation is anything but automatic. There's still the element of *speed*. You might boast bountiful brainstormers and make strong prototypes, but if you miss all your deadlines and ship dates, everything you've worked toward will be for naught. It's not enough to master the basics. You must master them in prime time, under the stress of competition, under the glare of a client losing patience. We get to practice this every day, but even that's not enough. Though we're very good at working fast, sometimes we need a refresher course. A race to see if we're as good and fast as we'd like to be. It's an essential element to keep a team or a company in tune. Here's the story of one of our races. Perhaps there's a competition that might keep your company "battle-ready."

Over the last few years, the Sand Hill Challenge has become our Indy 500, Kentucky Derby, and America's Cup rolled into one. It's a soapbox derby put on by Silicon Valley venture capitalists to raise money for a charity called Safe Rides, which offers kids a safe, anonymous ride home if they've been drinking or hanging out with other kids who might not be sober. Dave Lyons, then head of our machine shop, got us excited about the first Sand Hill Challenge, which took place in September 1997. It seemed a no-lose proposition. The cause was a good one, and what better way to challenge our prototyping machine than to build a gravity-powered, aerodynamic race car.

Our backer was the venture capital firm of Mohr Davidow, who figured if we could design laptops and computer mice, why not a soapbox speedster? We've always believed that our experience and innovation

methods make us a quick study. But there was something else here. The race gave us a chance to test ourselves against some formidable competition.

PLAY IMITATING WORK

Lyons pooled the efforts of IDEO volunteers, held some quick brain-stormers, and within a couple of weeks our first prototype was barrel-ing down the hill. When you're short on time, it's critical to make your dumb mistakes right away. After cobbling together a couple of three-wheelers, we finally got around to reading the rule book and realized we were a wheel short. Our early cars were pretty crude (many were thrown together out of old bike parts) and not terribly safe. Nearly everybody was afraid to test-drive, so Lyons volunteered, but even he got a bit queasy when he topped thirty-five miles an hour in the bike lane and started wobbling toward traffic on not much more than a sheet of plywood with wheels.

This soapbox derby was more challenging than most because each team was allowed a couple of pushers. Our lead muscle was former San Francisco 49ers running back Roger Craig. Each Saturday, the team would meet at a parking lot by Sand Hill Road at daybreak. Why so early? The racecourse ran through a major traffic light, and the chances of a collision were lowest in the early morning. Craig and an-other ex-football player, Scott Davies, gave the car a good running heave, and Lyons would cross his fingers and hope for the best. Four IDEOers took command of the intersection, stopping traffic for the flying fifty-pound soapbox and its intrepid test pilot. The team would race for hours, making adjustments and checking results. Sometimes the cops would tell them to pack up and go home; sometimes they'd just tell them to be safe and stay in the bike lane.

Every element of the car was rigorously prototyped and tested.

Nothing was as simple as it seemed. Take pushing the car. Because of our car's teardrop shape, there was nowhere to push. So in a back alley behind our shop the team tested all sorts of approaches. They'd brainstorm for half an hour and build the concepts an hour later. They must have made eight or nine pushing mechanisms, with the best solution boasting nearly as much design as the car. The design evolved into a T-shaped frame with Rollerblade wheels that docked to the back of the car but stayed behind as the car sped away.

At last, the race drew near. The entrants arrived the day before the contest like kids at the beach, showing off their muscles. By this time, we had tested cars that steered in the front as well as the back, cars that were heavy, cars that were aerodynamic, and cars that were light. We found some great wheels by studying the buggy race, a similar contest held at David's undergraduate alma mater, Carnegie Mellon. A couple of IDEO Carnegie Mellon alums described how they used wheelchair racing wheels, and we tracked down a set from a small catalog company that specializes in sports equipment for athletes with disabilities. But the more cars we built, the more we discovered tradeoffs. The heavy cars, for instance, reached the highest speeds, but they accelerated more slowly.

Another successful venture capitalist had hired a talented engineer and race car builder, known for building antique race cars for the Laguna Seca races. The car arrived on a shiny race car trailer. It was huge, a retrofitted Formula race car, stuffed with stacks of weights. IDEO's entry wasn't nearly so impressive. Ours arrived on a rusty old flatbed trailer, the car little more than a sheet of plywood sheathed in a teardrop-shaped carbon fiber shell. The pre-race trash talking started ramping up. The celebrated car designer ambled over and informed Lyons he'd obviously failed physics. Lyons didn't know what to say.

"You haven't done the numbers," Racing Man lectured. "The heavy car is going to win *big*," he said. He boasted that his car had clocked sixty-five going down the hill. Lyons swallowed hard. IDEO's car hadn't done more than forty miles an hour. He worried that maybe the race was over.

The next morning some twenty thousand fans lined the course. It was tense right from the start. The temperature was close to a hundred, and to increase the aerodynamics, we literally taped the bubble over the driver, who lay prone, just inches from the ground. We were afraid our driver—IDEO engineer Amy Han—might pass out with the long wait between heats. So as we waited for the starting gun, we stood over the car with a couple of reflecting space blankets, blocking the sun. And then suddenly it was our turn. Roger Craig and Scott Davies gave the car a mighty heave, the T-bar fell back like a rocket ship's propulsion unit, and IDEO's lightweight plywood and carbon fiber car leaped off the line.

Meanwhile, the converted Formula race car started like a tortoise. Oh, it was pushing sixty at the finish—but IDEO won going away, beating it by a dozen car lengths. Even better, the Sand Hill Challenge raised $100,000 for the Safe Rides program. That's about as good as it gets. Play and adventure imitating work—all in the name of a good cause.

To me, the Sand Hill Challenge is not just a great story of teamwork but even more a lesson in the value of quick prototyping. Racing Man may have mastered his physics, but he didn't spend enough Saturday mornings testing to learn that when it comes to soapbox derbies, there's more than one way to get to the finish line first.

ONE-HIT WONDERS

History is full of one-hit wonders, and the next year we knew it wasn't going to be easy to defend our title. NASA, Lockheed, and the designer of the Voyager—the plane that circled the globe without refueling—were building cars. How could we compete against such a stellar group? We started the way we often launch an IDEO project. We don't

pretend to know all the answers on day one. We looked for help and expertise.

We knew from the first race that wheels were the biggest factor. We did some research and learned that Michelin had built special wheels for solar race cars that greatly reduced resistance. A couple of guys from the Stanford Solar Car team helped us find these tires. The real challenge turned out to be designing and building a rim for them. With a little more detective work we located a document that drew the profile of the rim, and worked with a nearby company that makes cooking woks (honestly) to turn out custom-spun aluminum rims.

Next, we tackled the human equation. Roger Craig and Scott Davies were great football players, but they weren't necessarily the best race car pushers. Who would be among the best in the world?

With a lot of luck and some help from our partners at Mohr Davidow, we met Darrin and Dan Steele, fresh from their experience with the U.S. Olympic bobsled team in Nagano, Japan. The Steele twins, bobsledders by winter, decathletes by summer, were big and strong and excited at the prospect of pushing and helping to design the car.

We were playing with some intriguing new ideas. Certainly, the trash-talking race car designer's idea that a heavy car would go faster was right—but only if you could somehow get off the line fast. Could we combine the best of both designs—a light car at the start and a heavy car after letting go?

Bobsledders were the answer, of course. That's what they excel at, pushing and then hopping on the car. So we built a car to do a controlled experiment. We outfitted the car with an accelerometer and visited a guy who owned a go-cart factory. The project was contagious and ridiculously fun. The go-cart guy started showing up several nights a week, offering ideas and outfitting our car with advanced Grand Prix data-logging instrumentation.

But our test surprised everyone. When the bobsledders leaped into the car, it actually made it slower than our one-person version. The dif-

ference? When they hopped aboard, the energy of that last bench-press shove was lost.

So how could we capture the best of both—the last heave and the added speed of the heavier car? We tried having one of the twins hop on board while the other shoved him down the hill. Unfortunately, the slim car wasn't terribly stable. There was a minor crash and some serious road rash. The U.S. Bobsledding Association—and the sledders' girlfriends—advised us to return to our one-driver design.

Small crashes sometimes happen when you're trying to innovate. It's part of the territory. We've had this happen a hundred times with major products, a feature or design that wasn't quite ready for prime time. We didn't despair. We just kept plowing ahead with our one-driver design. The final race was close, but in the end, we edged out a team of master solar-car builders from U.C. Berkeley.

In two short years the Sand Hill Challenge had become an IDEO tradition, a matter of pride, and a tremendous creative outlet for many of our top people. But after winning two years running it didn't make sense to build another speedy car. Routine is the enemy of innovation. Do the same thing the same way too many times, and it's bound to become routine. It's why smart companies switch their sponsorships of athletic events and change the theme of their retreats. So in 1999 we entered a beautiful red Chinese dragon mounted on three engineless motorcycles, and won the "whimsical" category.

Not everyone had learned their lessons about the importance of prototyping. Another prominent Silicon Valley firm built a car with four wheels all in a line like a giant Rollerblade. They designed the car on the premise that aerodynamics were critical. But we knew from our Saturday morning tests that friction drag in the wheel was what counted. The cars didn't go fast enough for aerodynamics to be a big factor.

Nonetheless, the design firm focused single-mindedly on its in-line, aerodynamic car, much as we've seen companies prematurely rush consumer products into manufacturing. They spent hundreds of hours on it but apparently didn't leave time for a single test run.

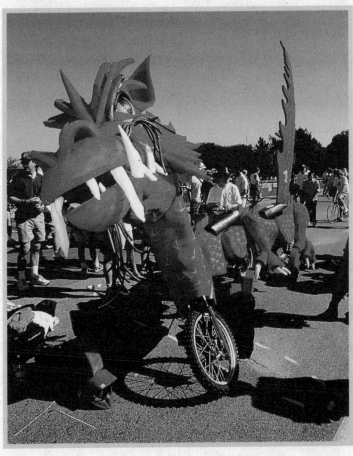

Joe Watson

Off-beat competitions can inspire teams.
The team had to rush to get its whimsical
red dragon ready for race day.

Come race day, the announcer was trumpeting how this world-famous firm had built such a sleek, amazing vehicle with a unique in-line wheel design. Like a great concept car, it was something to behold. The gun went off, the stylish soapbox looked absolutely marvelous for a few seconds, then wobbled and rolled onto its side, skidding to a stop.

The announcer deadpanned that he didn't think he'd ask them to design a car for him anytime soon.

GETTING THERE FASTER

Over the years we've found that entering the Sand Hill Challenge and other competitions have become a critical part of reinvigorating the IDEO Way.

Innovation doesn't happen in a vacuum. The faster you can come up with improvements or genuinely new products, the more likely you'll succeed in the marketplace. But you don't simply order up speed. And the path to superior innovation is rarely a direct line from A to B.

Look for competitions that you might enter (or create) for your teams that emphasize the playful, speedy side of innovation.

You just might get to the finish line faster.

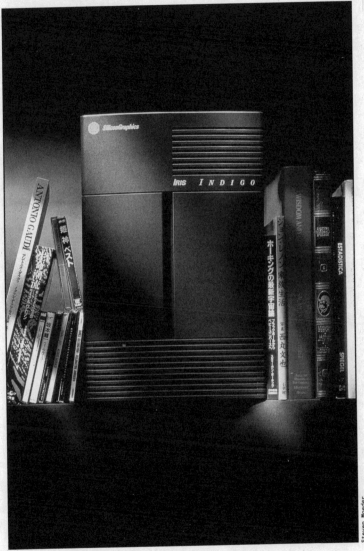

aking decisions about innovation means taking risks. Every executive and team leader faces this reality. For Silicon Graphics CEO Ed McCracken, it was a choice between staying with the status quo or breaking away from the pack. He told us loud and clear that his high-end graphics computers didn't *perform* like the beige computers made by everyone else. Why should they *look* like them?

IDEO's designers and Silicon Graphics' engineers realized the challenge was to create a palette of colors that would surprise but not shock. So was born the world's first purple computer. Silicon Graphics went on to grab the public's attention with its distinctive "Indigo" machine. Green, red, and blue iterations followed, making Silicon Graphics synonymous with bold color and irreverent playfulness.

Sound a little like the colorfully translucent iMac? When the iMac debuted in "Bondi blue" in 1998 and began appearing in different fruit flavors the following year, the idea seemed as novel and appropriate as Silicon Graphics' first splash of color almost a decade earlier. There's plenty more to the iMac story, of course, but the machine's bold shape and color undeniably helped fuel its success.

Risk takers. How can you know when and where to take the plunge? Sticking with the status quo always seems safer, even though it loses in the long run. Risk taking happens on the microlevel throughout a project. It's the big risks that we're unsure of. The ones that define what your product will and won't be.

Think of the countless Web ventures that have sprouted up over the last several years. The best have invariably been entirely new companies diving into the unknown. Grab nearly any slice of the Web, and you'll find a company taking a risk—Hotmail gambling that free e-mail could sell, Amazon trying to sell something as bulky and old-fashioned as books, iWon.com trying to grow rich in the long run by giving away millions of dollars to its customers in the short run. Now that the anx-

iety about using credit card numbers on the Internet has diminished a little, we may tend to forget that virtually all of the early business-to-consumer e-commerce sites took the risk that fearful customers would stay away in droves.

FAIL YOUR WAY TO SUCCESS

We've got a saying around IDEO: "Fail often to succeed sooner." Failure is the flip side of risk taking, and if you don't risk, odds are you won't succeed. It's a lesson we've learned the hard way, and it's one we've seen with countless clients.

Good companies embrace a culture of mini-failures. This goes to the heart of our prototyping method. By trying out features or capabilities in rough form as fast as possible, you've got a better chance of honing in on the critical fundamentals when the ship date rolls around. Nobody worth their salt hasn't failed on a big stage. Take enough chances and you risk a few big failures. Charles Schwab, for instance, failed a number of times before hitting on a successful formula for e-Schwab. Indeed, you could argue that the story of his business has been a series of risks, failures, corrections—and hits. Schwab himself claims that his company has made so many mistakes that they have coined the term "noble failures." Here are just a few of his company's screwups.

Pocketerm, a handheld stock-quote device, was overly expensive and bug-infested. Schwabline, a system that would provide paper copies of people's portfolios, had a bad habit of breaking down and could only provide information a day old. And Schwab Quotes struggled to make real-time quotes available for customers with a touch-tone trading system but was very complicated. Other Schwab systems or software failed because they were too cumbersome or buggy.

What's the lesson? All these "noble failures" kept Schwab on the cutting edge. By the mid-1990s as the Internet was heating up, the

company was ideally positioned to take its biggest and most successful risk, one I'll talk about later in the chapter.

"The noble failures always kept us pushing the barriers," says Schwab. "We kept believing that a mistake may not be a mistake when it reflects an essentially sound strategy."

Failure helps teach risk takers which rules can be broken and which are there to teach you a hard lesson. We've seen our share of failures. In the early 1980s, for instance, we decided to introduce a computer accessory, a clever monitor stand for the Apple II. Our product quality was good, our features strong, the price under forty dollars. Things were looking very good, except for one problem. Just as we went to market, Apple abruptly changed its mind and decided to create its own stripped-down version, which it shipped with every machine—for free. *Free* is pretty hard to compete with. What did we learn? Never ignore the worst-case scenario, in this case someone creating a business model that undercuts your pricing or renders your product unmarketable.

Neophytes tend to become infatuated with their product or business plans, and we were no exception. One of our biggest early lessons was an Italian love affair, a beautiful design that was a commercial failure of operatic proportions. In the early days of our firm, David got a chance to meet the revered Italian designer Ettore Sottsass. They really hit it off and decided to combine Sottsass's Italian styling with our Silicon Valley engineering in an electronic consumer product. The playful cubist red, black, and yellow Enorme phone created jointly with Sottsass Associati was a runaway aesthetic and cultural success. The phone entered the permanent collection of New York's Museum of Modern Art, appeared in movies, and was featured on the cover of European design magazines.

Sottsass came through with a striking design. How hard could the logistics of basic telephony be? In retrospect, trickier than we thought. Inexperienced at the bargaining table, we signed up an unproven manufacturer in Hong Kong and entered into an exclusive distribution agreement through a tiny, high-margin channel. We discovered that

Steven Moeder

The Enorme phone was a joyful failure; today
it's in the Museum of Modern Art.

the Enorme phone we made in Hong Kong for $25 would cost $150 at Neiman-Marcus and other upscale retailers. It didn't help that, around that same time, *Time* magazine started giving phones away with every subscription.

The Enorme telephone was one of IDEO's biggest and most joyful failures. I'd argue that every company or individual needs at least one. Sometimes you've got to take risks you don't understand, and maybe even lose some money if you want to gain buckets of experience.

So how can you survive your joyful failures? I think juggling is the perfect metaphor for intelligently embracing risk taking. Learning to juggle can help give you the right mind-set. Face it, we're all going to drop the ball once in a while. As I learned from my fellow Palo Altan John Cassidy, author of *Juggling for the Complete Klutz,* the trick to beginning juggling is not to use rocks, which may bruise your feet, or balls, which tend to roll under the couch. Beanbags are ideal for learning to juggle (and Klutz Press ships three with each book) because they put such a small price on failure. If you start with tennis balls, they'll quickly roll away, and you'll spend so much time scampering after your errant juggles that you'll likely quit after your first few tries. Beanbags encourage you to keep juggling until you get good enough at it to move up to balls—or knives, for that matter.

In the business world, of course, you can "juggle beanbags" by doing lots of prototyping in inexpensive, easily malleable materials, by running ideas and rough presentations by bosses or clients early on in stages. That way when prime time approaches and the real deadline looms, you'll be confident enough to juggle knives.

GROWN MEN AFRAID OF A LITTLE MOUSE

Sitting on the sidelines, waiting for the "riskless breakthrough" or the "ideal risk" to come along, doesn't work. Those who try to scientifically

measure opportunities sometimes miss the boat. For example, in the early 1980s we discovered that the computer mouse no longer really needed its electrical "tail." Pop in some Nicad batteries, add an infrared transmitter, and you've got a cordless "mouse on the go."

We proposed the innovation to a client at a very large corporation. He loved it—at first—but the next day, fear set in. We've seen it before. Everyone fears taking risks, but at bigger companies the penalty for risk taking is often larger. We weren't that surprised when a couple of weeks later the formerly enthusiastic manager turned timid. "If this fails," he told us, "I'll be known for the rest of my career with the company as the guy with that stupid cordless mouse."

Instead, he took the easy way out. He passed on the innovation and settled for a bland product. A couple of years later we got the chance to create a cordless mouse for veteran Silicon Valley risk taker David Liddle and his dramatic Metaphor workstation. Metaphor got a lot of press attention for its innovative new product and proved that the technology could work successfully.

MAKING NEW RIDES

Thinking about what you might lose—market share, revenue, a title, your status, your job—makes it nearly impossible to take a leap. That's why the big new ideas usually come from small companies, or big ones who somehow manage to act small. Fear doesn't get you down the mountain.

Take the amazing phenomenon of snowboarding. Plenty of journalists and ski resorts at first called it a fad. During the early ramp-up of snowboarding, *Time* dubbed it America's "Worst New Sport." Ski resorts tried to ban boarders, and mainstream ski manufacturers figured it was a passing fancy.

But looking back, it was the ideal time to break the rules. Downhill skiing had been stagnant for years and was starting to lose its momen-

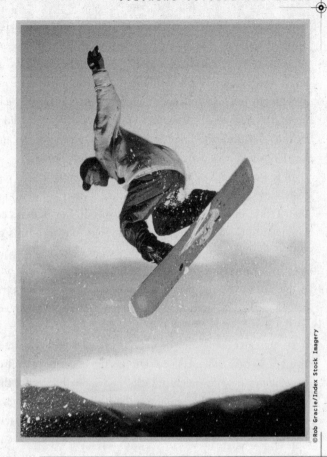

©Rob Gracie/Index Stock Imagery

Snowboards reinvented the ski industry.

tum as rising ticket prices were keeping families and children off the slopes. Once a symbol of youthful rebelliousness (remember Robert Redford in *Downhill Racer*?), skiing was becoming dated.

Great time for the dominant downhill ski makers to reinvent the sport, right? Not exactly. Rossignol, Salomon, Fischer, and the other ski powerhouses were too busy protecting market share in their mainstream business to ride the first snowboarding wave.

Meanwhile, upstart Jake Burton Carpenter was breaking the ice—

so to speak—by hacking out boards in his Vermont garage. He went to his local hardware store—and Austria—to get ideas. In the early days, Burton's toll-free customer service line rang in his bedroom at all hours. He built one of the first boards that could handle ice and hard-pack, partly because Vermont's conditions were so icy. While he didn't invent the snowboard (being first isn't always critical), he did popularize the sport through a combination of quality, persistence, and lobbying that would have done Macintosh evangelists proud.

In retrospect, it's easy to see why manufacturers and ski resorts initially fought the tide. Early boarders had a nasty habit of embodying the worst of male adolescence at high speeds. But Burton's campaign to convince ski resorts to open their mountains to boards gradually took hold. Within a few years, 90 percent of U.S. resorts accepted snowboards, and now over 5 million Americans enjoy the sport. Burton is the world's leading manufacturer of snowboards, and even IDEO has hitched a ride on the sport, designing face-hugging goggles for use on the slopes.

It's the same story when it comes to biking down a mountain, an activity we know firsthand, since IDEO is full of cycling enthusiasts that have helped develop helmets, locks, and sports bottles for the mountain bike industry. Who would you think blew an inside track to "own" mountain biking? Schwinn, one of the biggest bike manufacturers, of course.

Believe it or not, there was a time when the company was ideally positioned to dominate mountain bikes.

Schwinn succeeded wildly in the sixties and seventies. In 1963 one of Schwinn's sales managers noticed California kids souping up bikes with banana seats and Texas longhorn handlebars. A product design manager whipped out a prototype Sting-Ray cycle, and soon the chubby-tired bike was the company's best seller.

At its zenith, it sold a quarter of all U.S. bicycles. When mountain bikes emerged, however, Schwinn mistakenly wrote them off as a fad. That despite the fact that some of the early mountain bikers—just north of us in Marin County, California—had first tackled the rugged

mountain paths with souped-up Schwinns, partly because the bikes were so darned sturdy. Success has a tendency to hamstring companies, to cause them to become complacent and neglect the risk taking that got them to the top. Though Schwinn adeptly picked up on the customer-inspired Sting-Ray trend in the sixties, somehow the company managed to miss out on an analogous mountain biking trend in the eighties.

When the dust settled, of course, mountain bikes proved to be much more than a fad. The public flocked to the fat-tired, easy-to-ride bikes for lots of good reasons. Ten-speed road bikes, formerly the most popular style, were actually built for racers, not recreational use. Thin tires may be fast, but they aren't comfortable, convenient, or especially safe. And, of course, there's the beauty and thrill of riding down a mountain. We all know that some of those mountain bike tires never even touch dirt, let alone steep mountain trails, but mountain bikes are far better suited for the average rider than the old low-slung ten-speed.

Specialized Bicycle Components, for whom we helped design biking accessories, didn't have the problem of resting on its laurels. Originally a niche vendor of high-end bike parts, Specialized quickly capitalized on the mountain bike craze, popularizing the sport with races and clever promotions, while the old U.S. giant, Schwinn, slowly became irrelevant. Today an amazing 70 percent of the full-sized cycles sold in the United States are mountain bikes. Based on its success as a mountain bike maverick, Specialized still uses its in-your-face slogan, "Innovate or Die," which is printed in huge white letters on the long black hearse the company takes to biking events. Not a bad motto to take to heart.

SPIRITED SKIES

As the quickly changing sports equipment landscape emphasizes, risk taking is a moving target. If you don't take some chances, you get left

behind. But does that approach work for every business? What if your industry prides itself on tradition and safety? What about a more staid field like the airline business?

During my years as a management consultant I studied many of the world's airlines and wrote dense reports about how they might improve their operations. I always admired Southwest, not only for its humor but also for its spontaneity. Legend has it that Southwest was launched when CEO Herb Kelleher drew up the airline's business plan on a cocktail napkin. You won't find those boring standard safety lectures required by the FAA on Southwest, where the easygoing flight attendants embellish the instructions with little asides like "secure yourself around the nose and mouth and breathe like you've never breathed before." The jokes don't just get chuckles out of passengers, they get them back on Southwest flights.

To me, that's the best kind of rule to break—an established pattern that does nobody any good. Indeed, even the stodgy FAA chuckles at Southwest's jokes, noting that anything that gets passengers to listen to safety briefings is fine by the agency. Southwest's funny and friendly flight attendants have become as much a part of the Southwest experience as the planes. On a long taxi to the gate, now you almost expect a couple of corny jokes over the intercom.

But cracking funny bones at thirty thousand feet is just one of the ways Southwest bypassed the traditional rules of running an airline. Kelleher brought his intuitive style to what used to be considered an impersonal industry. Southwest is renowned for personal touches, like sending birthday cards to every employee. Kelleher says his only rule is making as few rules as possible, an antimanagement style that seems to resonate with workers: *Fortune* recently named Southwest as the best company to work for in America. The good vibes have removed barriers that used to exist between classes of workers, with the result that Southwest topped the rankings in safety, price, on-time performance, baggage handling, and customer service. Things happen faster at Southwest because workers don't get stuck in thinking that they can

only do things that fit their job description. The gap between tasks has narrowed. Everybody works as a team. Flight attendants (and occasionally even pilots) tidy up before deplaning. Ticket agents carry luggage in a pinch. The focus is on solutions, not who might have slipped up. And the good vibes seem to infect passengers too, who seem more tolerant and friendly than those on more "corporate" airlines.

Not surprisingly, this can-do spirit has led Southwest to upend other stuffy airline routines. For instance, by making seating first-come first-served instead of assigned, people can actually choose whom they sit next to (and the colored boarding card simplifies boarding for gate agents). Southwest is great at short-circuiting convention. The airline abandoned that aisle-clogging drink cart by taking your order before takeoff. Southwest seems to do well on the Web too, with a simplified, "fewer clicks" approach that has resulted in a high "look-to-book" ratio.

BEING BIG AND ACTING SMALL

Southwest may have grown into a major airline, but it hasn't lost its responsive, entrepreneurial culture. Thinking and acting small is what counts. This is true regardless of the company. I'm constantly surprised at how willing some of our large clients are to take calculated gambles.

We've worked a lot with Procter & Gamble over the years. The $38-billion consumer products powerhouse is widely praised for its marketing muscle and R&D.

But that size was likely to get in the way when Internet opportunity came knocking. So a handful of P&Gers flew to Silicon Valley and got a crash course on the new medium. They teamed up with the venture firm IVP and spun out a separate entity, Reflect.com, free of the normal P&G hierarchy. Reflect.com's business plan seemed radical for a brand-conscious, traditional firm like P&G. The goal was to offer phe-

nomenal personalization—letting women mix their own products online. I wouldn't dare guess whether it will be a success, but the effort shows that you've got to experiment with new formulas, even when they appear to run counter to your mainstream business. In the bricks-and-mortar world, P&G, for instance, can sell only the top dozen or so shampoos. Reflect.com, meanwhile, will be able to offer tens of thousands of unique, individually customized products.

Not only is Procter & Gamble taking a big corporate risk, but, true to Reflect.com's personalization philosophy, the company has also put risk at the individual level. Procter & Gamble managers joining the Reflect.com team must actually quit their P&G positions and work "without a safety net." Each individual's success is tied to Reflect.com's performance.

Why should a company that dominates a market bother with risks? Plenty have preferred to stick to old ways and been caught flat-footed by nimble new competitors or novel technologies. Not Gillette. Having just spent over $200 million creating its successful two-bladed Sensor razor, the company poured phenomenal resources into its new Mach 3. Seven years in the making, the new razor consumed an unheard-of $750 million in development and manufacturing costs. Innovations included the extraordinarily thin, stronger-than-steel blade and, of course, the three-bladed system. Within months the Mach 3 was outselling Gillette's previous front-runner, the SensorExcel, by an incredible ten-to-one margin. The company's dollar share of the razor blade market topped 70 percent, its highest in nearly forty years. Pretty aggressive behavior for a company that could have just rested on its laurels for a few years. A skeptic might say that they cannibalized their own market share, but in the process they reestablished themselves as the industry leader and trendsetter.

BREAKING WORK

My brother David hates rules. He hates them because he knows that when you start making rules, you sow the first seeds of bureaucracy.

We reject titles and big offices because they pose mental and physical barriers between teams and individuals. Paperwork is kept to a minimum. There's a friendly competition at work in the company. People grab projects and opportunities.

Not long ago, one of our young designers started an office in Germany. Now, it made perfect sense: Europe was ripe for an IDEO office and we had the right connections. But I had to chuckle: The designer was already on-site in Munich with a makeshift office space and his own computer equipment before he had written a proposal or asked anyone's permission. He certainly hadn't filed out a "New Foreign Office" requisition form. He'd just announced that he was doing it, and no one could come up with a good reason for slowing his momentum.

The biggest rule breakers change our processes. They literally change how we work. Sometimes failure can be the inspiration. Who would imagine that being a mediocre typist could make you a millionaire? But for Bette Nesmith, an executive secretary at the Texas Bank & Trust in Dallas in 1951, it was an opportunity waiting to be exploited. Nesmith learned something critical while helping decorate the holiday windows at the bank. The artists corrected their mistakes by painting over the error. Nesmith decided to try the same at work, a telling example of how observation can lead to successful cross-pollination.

Some considered her little swipes of tempera water-based paint "cheating," and one boss warned her not to use the "white stuff" on his letters. But soon Nesmith had a cottage business supplying Mistake Out to fellow secretaries. She renamed the fast-drying, nondetectable fluid Liquid Paper and applied for a patent and trademark. IBM wasn't interested in buying her business, so Nesmith plowed ahead alone.

Within a decade she sold out to Gillette for nearly $50 million—plus royalties.

Those of us who remember the days of typewriting can thank Nesmith for making it possible to finish all those college papers without having to retype the whole page. Nesmith changed how business documents were created. It's a classic story with a modern connection.

LINUX BREAKING THE RULES

For decades, Microsoft, IBM, and even idealistic Apple dictated the way software was developed and protected. Software was always considered intellectual property—as real and tangible as a car or an airplane—and you'd no more give away the code to your new software than you would the combination to your bank vault.

Then, in the early 1990s, Linus Torvalds, a twenty-one-year-old student at the University of Helsinki, burst onto the scene. Torvalds couldn't afford an operating system, so he spun out his own. The young programmer broke countless rules in the process of creating Linux. He never asked permission, never wrote anyone to see if it was feasible, never worked up a marketing or business plan, never estimated how long it might take—or, for that matter, thought about making money. Some of the best rule breakers aren't in it to make profit. They have a point of view, and little sense of the enormity of their goals. Torvalds wanted to build a cool operating system for the whole world. He wanted it to be free. He wanted a lot of help in improving it. So he made it "open source" software. Software freely available to the world. Software that could be improved upon by thousands of other talented programmers interested in improving it.

A professor with a competing operating system philosophy told Torvalds that his plan was in "error" and that if he were in his class he

"would not get a high grade for such a design." A cocky Torvalds snapped back in e-mail, "That's OK. Einstein got lousy grades in math and physics."

Linux has succeeded beyond Torvalds's wildest dreams. Today the Linux operating system has some 15 million users worldwide, and thousands of independent developers are working to make the system better. Major corporations initially wrote off open source software as fringe, unreliable products. But as Linux has continued to get stronger, one by one the corporate competitors have decided to join the movement rather than fight it.

This programming revolution is challenging the old assumptions that the best software is proprietary and maintained by a single company. Linux, the popular Apache Web server, and the widely adopted Perl scripting language have proved the advantage of having thousands of self-motivated programmers.

GIVE AND RECEIVE

The Web has sprouted plenty of smaller versions of the open source method. Think of the countless Websites that offer customers something free—search engines, greeting cards, e-mail, reviews, articles, Website storage. E-commerce is still in its infancy, and I'm not foolish enough to guess who will rise and who will fall. But it's abundantly clear that many of the basic tenets of business have been broken. Onsale, for instance, one of the first Web auctioneers, pursued what seemed at the time the prudent strategy for the Net's early technical bias, mainly selling blocks of computers to large corporate buyers. Christie's, Sotheby's, and Butterfield's, after all, proved that auctions weren't about the masses. But eBay flipped that dynamic on its head, realizing that if anyone could sell anything to anybody, the upstart

would have the advantage of mass and momentum. Onsale never grew because it wasn't a place for the average person to sell anything, while eBay rocketed ahead on turning the world into one big on-line swap meet.

COLORING OUTSIDE THE LINES

There's a certain serendipity in breaking rules. You can't always predict where an innovation might come from. It isn't something likely to spring from commissioning a report or doing focus groups. Sometimes it's a subtle mind-set. Growing up, I thought doing a "perfect job" in a coloring book meant placing one color within each space, being careful not to let it spill over a line. Meanwhile, my brother David splashed colors wherever he pleased, adding highlights and shading to create something that went way beyond a color-by-numbers approach. That's how rule breaking starts.

The key is taking a fresh approach. Take the time a hot East Coast advertising firm hired us to help it win a major account with one of America's leading mattress companies. They took a risk—and so did we. We brought in Skyline, our toy invention division, to lead us in brainstorming ideas for an in-store mattress demo. What would toy wonks know about mattresses? Not necessarily a heck of a lot, but Brendan Boyle and his team are old hands at brainstorming up quick, inexpensive demonstrations of toys. Why couldn't they dream up something for demonstrating mattresses? Sure enough, in a couple of days, Brendan came up with a clever talking pillow that described the mattresses' features, and the ad agency won their big-name client.

Brendan's team gets called upon frequently by IDEO because he's an old pro at shaking new ideas out when even our best teams get stuck. Brendan understands how to strip processes down to their basics, vary the mix, try new things.

Since Brendan is constantly pitching new toy ideas to major toy companies, he can't afford to get in a presentation rut. Formulas don't fly. Sometimes he'll whip up a thirty-second video, capture an idea on an art board, work up a quick prototype, tap something out on a laptop. It's a combination of what seems right for the idea, a blend of experience and intuition. His method is not falling in love with the tool. He isn't interested in spending a lot of time in video production or fine-tuning a graphics file. He might boost up a hand drawing in the computer, even add a little rendering to tell a story. Sometimes he scratches out a flurry of sketches on yellow legal pads and hands the ideas off to illustrators to draw them up. You can't limit where or how ideas get presented or expressed. Coffee breaks at Skyline can be brainstorms. Brendan has even taken his team to a nearby range to hit some golf balls before sitting down at a picnic table for a brainstorm. The person who toils endlessly at his desk is not likely the person who is going to hatch a great innovation.

Rule breakers are found in all sorts of industries. Sometimes innovative ideas are spawned in fields or product categories that normally discourage innovation. Here's a short list of seven surprising rule-cracking companies and their products.

IGLOO

In most compact refrigerators, the main purpose of the freezer compartment seems to be to spawn miniature glaciers. Igloo asked college kids what they'd like (more room for six-packs!) and started chipping away at convention. Igloo turned to the latest thermoelectric technology, used primarily for ice chests. The problem was that Igloo designers weren't sure it could cool a two-cubic-foot fridge. But by canning the freezer, the new technology was not only plenty cool enough, it also fit one and a half six-packs more in the same size fridge. And so the Space Mate was born.

I don't know about you, but I think nine out of ten college guys would agree: More cold beer in the same space is a genuine innovation.

SHOEBOX GREETINGS

Imagine if you had to turn out twenty new ideas a day. That's the goal at Hallmark's top-producing humor division. They do it by eschewing convention. No rules about dress or office decor. Writers are free to wander the halls and even brainstorm outside when the muse strikes them. Goofing off—flipping through magazines or watching videos—is encouraged, and often, writers will be handed afternoon movie passes. Writers and artists who need an extra charge can periodically slip off to company art studios and dabble in woodworking, ceramics, and other creative pursuits. This encouraging approach to creativity hasn't just worked in the marketplace of ideas. Employee turnover at Shoebox is under 6 percent.

SEPHORA

We can learn a lot from companies that retire old retail routines. Sephora, for example, reinvented the way high-end cosmetics are sold, ending the reign of white-frocked commission-based saleswomen. The French-owned company brought expensive perfumes and cosmetics out from under the glass and offered dizzying arrays of samples. Today department stores are rushing to copy the upstart. Instead of the branded "sales push" approach used almost universally at department store cosmetic counters, Sephora groups products by category and encourages women to browse at their leisure. Early results suggest they're actually buying more without the push.

TARGET

Bold companies change their colors. Target, once known primarily for its low prices, has now become a design leader. Teaming up with famed architect Michael Graves, Target introduced a whimsical housewares line. The wildly successful Graves Collection proved contagious and eventually included some two hundred items in various

product categories, such as garden furniture and timepieces. More than just introducing design for the masses, Target was determined to break out of the decades-old idea of a discount store.

SWATCH

Even the most ardent classicists can up the tempo. Swiss watchmakers, for example, countered a Japanese digital timepiece invasion through corporate jujitsu. They halved the number of working parts, introduced a quartz-powered analog watch priced far below anything previously sold in Switzerland—and led a design and fashion craze.

E-SCHWAB

Sometimes you've just got to face your losses and opportunities head-on. Schwab skirted bureaucracy and turned a routine internal automation project into a Web crusade. When e-Schwab premiered in 1996, on-line accounts boomed. But the new business proved a two-edged sword: Schwab brokers complained about the low on-line flat fee. Risking internal dissent and Wall Street's wrath, Schwab spread the same low Web trade commission to all of its customers. The gamble paid off. Schwab's risk taking arguably saved the company from being passed up, while shooting it ahead of traditional high-end firms like Merrill Lynch, which fretted for years over how the Web would affect its brokers.

RUBBERMAID

New product development is a 365-day-a-year process. Few of us would think of routine products for the home as being ripe for innovation. All the more reason to commend Rubbermaid's risk-taking approach to products. Whereas other brand leaders might ride their marketing clout, this Wooster, Ohio, company (just down the road

Steven Moeder

The Swatch redefined watches as a fashion accessory.

from where I grew up) sets itself a goal of introducing three hundred new items every year. Nearly a third of these are entirely new products, designed from the ground up. Routinely named one of the most admired U.S. corporations, Rubbermaid proves that innovation isn't always about technology or hipness.

NOT *TOO* FAR OUT

One caveat about coloring outside the lines: You have to constantly evaluate what's *too* far outside. The late Gordon McKenzie, a fellow member of the Innovation Network, would call this balancing act "orbiting the giant hairball." He spoke of how most companies have a con-

fining network of rules, regulations, policies, and mind-sets that constitute a messy hairball you don't want to get stuck in. On the other hand, if you just fly off into outer space, pursuing a completely independent trajectory, you may lose your ability to help the company or its customers or even yourself. The ultimate goal is to maintain a kind of continuous orbit, influencing the organization and being influenced by it, without ever getting completely snagged in the morass.

So go ahead and color outside the lines, but try your best to stay on the same page.

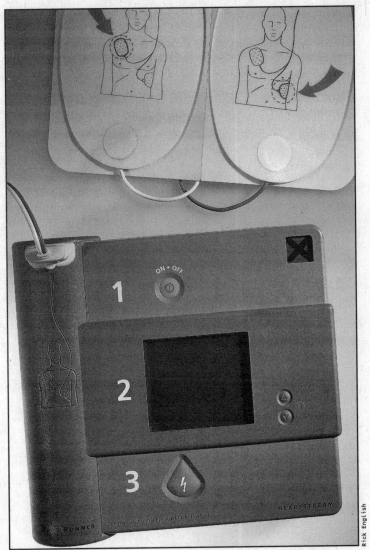

Rick English

ow do you refine an innovation? How do you streamline a product or service to the point where it's as simple and powerful as the Palm V or the Heartstream defibrillator?

It's no easy task, but if you're successful, you may just get a chance to come out with version 2.0. There's no finish line. Once customers start to use your products or services, you need to refine them again.

FIGHTING FEATURE CREEP

It can happen at nearly any point in the development or refinement of a product or service. It's the enemy of innovation, and every team molding a product or shaping a new service fights it sooner or later. When real innovation stalls, feature creep often sets in. Several years back, for instance, Matsushita asked us to help it introduce a new steam iron. What could be simpler, right? Think again. In Japan steam irons were laden with features that, while impressive, weren't really valued by their customers. A typical iron at the time had three different kinds of steam, a detachable water reservoir, and a little "garage" for storing the iron during cool-down. Packing those features in pumped up the price of Japanese irons to a whopping two hundred dollars, while the Southeast Asian models were hitting the market at thirty dollars. Ironically, many of the Japanese irons' more technical features didn't appeal to *women*, who made up nearly 100 percent of the domestic market. So we dumped the techno focus, eliminated features that were confusing or underused, and ushered in a simpler, softer iron that could be made for less than half the original cost.

We're all in search of what Carl Ledbetter, former president of AT&T Consumer Products, used to call the Wet Nap Interface, prod-

ucts built to be used as simply as the directions on those little moist towelettes you get with take-out foods: "Tear open and use."

We haven't quite scored the Wet Nap Interface yet, but we've tried. When we were designing Apple's first mouse in 1982, it was up in the air as to whether it would be better to make it a one- or two-button mouse. We believe Apple made the right decision in sticking to a single button. Apple was already asking people to shift from a world of keyboard-based commands to a graphic, mouse-controlled environment. Adding an extra button was too much to demand of users in the first version of the product.

I've seen feature creep happen so often that I can almost see it coming. When design is done by committee or the goal is fuzzy, you can be sure that feature creep isn't far behind. It's a little like restaurants that list dozens of entrées on the menu when only one or two ingredients in them are different. Remember how the original VCR was absurdly difficult to program? Nobody bothered to separate the unnecessary features from the critical ones, let alone create a user-friendly interface. If you find yourself struggling with a manual, you're likely a victim of featuritis. Texas Instruments, for example, once came out with a digital watch that practically required programming skills to set the time. Not surprisingly, it flopped. The thick manual was a sign that something was wrong. And we've all cursed overly complex digital clocks on the bedside table of our hotel room, when what we really want is something that works as simply as the old mechanical versions.

Everybody occasionally lets the details crowd out the objective. Who hasn't written a report a dozen pages too long or padded a PowerPoint presentation? Over the years, IDEO has received its share of run-on résumés that seemed to chronicle nearly every major life event. Not that we're perfect either. We once tinkered incessantly on a programmable robot for kids until we weighed it down with so many capabilities that it was too complex to understand and too expensive to sell.

We've all seen featuritis run amok. Open up the average in-flight mail-order catalog, and you'll probably find a cornucopia of feature-crammed products that seem to have no reason for being other than

the fact that an awful lot of people must have big attics. Most of us could live without an electric tie rack or a collapsible shoe horn.

The software industry has always competed by comparing feature lists, and as a result, the "new, new" version is always bigger and often clumsier than the one before. During the 1990s, for example, Microsoft introduced a series of new versions of its popular Word program. Each one seemed to be loaded with new features, but ran noticeably slower on some machines. (Imagine if Intel came out with a new chip family that was slower than the previous one.) Since I seldom use those high-end features, I feel like I'm driving a Chevy Suburban down a bike path.

Companies and, yes, executives, too, need to spend more time "driving" their own products. A few years ago, when Hertz introduced its NeverLost GPS navigational system, I became an enthusiastic early adopter. I'm dyslexic at map-reading and hopeless at directions. I like nothing better than to program in my destination and then follow the little arrows. But in the first two versions of NeverLost, the most basic feature was missing. Inexplicably, NeverLost lacked "a last-number redial," as in a "take me back where I just came from" command. If I've just driven from my hotel to the client's office, shouldn't there be a "return to origin" feature—like HOME on your Web browser? Last I checked, it took half a dozen clicks with a mandatory wait after each one on NeverLost to accomplish this.

Try "driving" your own products and services, as if you were trying them for the first time. My guess is you'll find something missing you may want to incorporate in version 2.0.

GIVE SIMPLE DIRECTIONS

Simplicity in design is like clear directions. Give customers fewer turns in the road, and they're more likely to lock onto their destination.

Where there's featuritis—or fat—opportunity is nearby. What is Sun Microsystems's Java software but the promise of a slimmer, simpler programming environment? Even the VCR, after a few years, became a little easier for taping one's favorite shows. Stereo and TV manufacturers, too, seem to have learned some of the same lessons experts put in practice long ago in fighter cockpits. Simplicity is essential. We can only remember seven numbers easily, and we can only make sense of a limited number of controls. That's why manufacturers created that little door at the bottom of your TV to hide the knobs you rarely use. Good remote controls also have little covers that protect us from obscure features, since we mostly use only the power, volume control, and channel buttons.

The first step to refining products or services is chucking preconceptions about how complex something needs to be. Virtually every industry has established thinking about which features are or are not necessary. To refine a product, you need to challenge that status quo. You need to be an editor, willing to cut away what's not essential to your story line.

Medical products have been notoriously feature-laden. But in dozens of products, we've managed to cut through that unnecessary complexity. My favorite of all the four thousand products IDEO has worked on is the sophisticated but simple Heartstream defibrillator. In spite of its complex underlying technology, the 1-2-3 interface on this device is so simple that even a child can figure it out. Literally. I handed my six-year-old a demo unit and asked her to give it a try. Without hints or help from me, she successfully pushed the big green start button and plugged the chunky yellow connector cords (which deliver the shock) into the defibrillator. (We were running it in training mode, of course, so the machine couldn't actually deliver a shock.)

"Now push the orange button," a calm voice directed my daughter. The same words appeared on a screen. The orange button flashed and beeped. My daughter pushed. In reality those steps could have saved a life.

We've learned that the public and even supposed experts are often

not the best judge of which features need to disappear. Refining products is not a popularity contest. You have to take risks and you will alienate some people. For example, a number of reviewers criticized Steve Jobs for not including floppy drives in iMacs and iBooks. But they missed the point. The iMac was designed to catch the Internet wave, and a floppy drive would have demanded extra space, fattened the price point, and focused on a soon-to-be-obsolete technology. Jobs wisely decided that the machine couldn't look like something from the Jetsons yet be weighed down with eighties technology.

SIMPLE AS A FRISBEE

At Stanford, where David teaches, he talks about "low mass" solutions. One of his favorite examples is the Frisbee, an innovation that bridged modern plastics and aerodynamics with the classic Greek discus. It boasts no moving parts, requires no instructions, and delivers fun with very little practice—all with about fifteen cents worth of molded plastic.

We've found that this simplicity factor applies to all kinds of products. Ironically, we've often found that version 2.0 can be an ideal time to streamline, but that's exactly the point at which software companies are tempted to add bells and whistles. We believe that approach is flawed and out of favor. Companies today understand that what people want is more integration and simplicity. The reality is that it takes time. You often can't achieve it in the first version of your product. You've got to get to market, gain market share. Version 2.0 may be your best bet.

Several years ago AT&T asked for some help in redesigning its phones. The telecommunications giant wanted us to create a "design language"—a kind of visual style guide—to help inspire and inform current and future designs. We eschew traditional "rule book" style guides and think you should too. They're often written too narrowly

and end up suppressing innovation. But AT&T had a huge problem on its hands. Decades of monopoly control had distanced it from customer needs and desires. When AT&T came to us, it had a mishmash of twenty-seven different handsets in active use. As Carl Ledbetter said at the time, "I don't know what the right number of handsets is for our product line, but I'm sure it's less than twenty-seven!"

The irony was that AT&T should have been designing great phones. They'd collected amazingly useful human data on how phones work best, everything from how to achieve an acoustic seal with the earpiece to optimizing the spaces between the keys. But in a company as large as AT&T, that knowledge didn't always transfer to the products.

We came up with something we called the Allegro design language. You've probably seen the phones it inspired, as the design language was widely implemented by AT&T and later by Lucent. Before, most AT&T phones were flat and rectangular with hard, off-putting edges. Allegro strikes a more human, receptive tone. The phone literally points to you, the receiver and handset tapering down and in. Keys are cupped to fit fingertips so you don't misdial. Small "signature details" help locate or emphasize functions. The answering machine's play button is the largest, for instance, and less-used features like flash, mute, and hold are grouped together. Color coding distinguishes between the telephony and answering machine functions.

HANDS ON

The best products understand something fundamental. You need to spend the most design attention at the place where you touch the product the most, like the spongy grip we made on the Oral-B toothbrush, so kids could "squish" the toothbrush in their fists, or the color-coded "touch points" on the McGaw Horizon infusion pump we designed that draw the nurse's eye to the points of interaction. We all

Jane Fulton Suri enthusiastically dived into the problem, examining nearly every element of people "mousing" in countless designed exercises that tested speed, comfort, and accuracy. The ball, for instance, had generally been positioned in the center or toward the mouse's rear. Jane proved empirically that moving the ball forward and letting people drive it with their fingers yielded greater precision—sort of like sprinting on your toes instead of your heels.

She also found that people naturally draped their fingers over the front of the mouse, suggesting that buttons should extend the width of the entire mouse, including the front. We made the left button larger than the right, because Jane's observations told us that people rarely used the right button. And we created a slight ridge running down the middle of the keys like the bumps dividing a freeway, to keep you from accidentally "chording"—clicking both buttons at once.

Credit for the mouse's comfortable-to-grip and "Dove Bar" shape should go to industrial designers Paul Bradley and Mike Nuttall. The Microsoft mouse was the first computer accessory to sport the glossy, self-cleaning finish originally found in telephones. Boldly white, it broke from previous beige and gray tones, and became widely emulated. The team also came up with what at the time was a novel branding solution. Simply printing the logo wouldn't do, since frequent handling would wear it down. Instead, we used a "double shot" process, in which the gray Microsoft logo was molded directly into the white plastic of the case. It worked perfectly, leaving an absolutely smooth surface and making the logo last for the life of the mouse.

The new mouse's look and feel made a huge difference, proving once again that design can make a breakthrough product. *PC Magazine* heralded "a dramatic, new shape" that was "a sheer pleasure to hold . . . the best feel of any mouse." The Microsoft mouse sold like hotcakes—more than 7 million units—providing Microsoft the brand recognition to expand into other lucrative hardware accessories.

HOW TO CREATE GREAT PRODUCTS AND SERVICES

Thousands of projects and hundreds of successful products have taught us some valuable lessons. Here are some of the objectives we try to keep in mind when starting work on the next new thing.

1. MAKE A GREAT ENTRANCE

Good buildings, Websites, and products make it easy to find your way about. They leave no doubt about which door to open or button to click. How do you welcome people to a service? When Wells Fargo asked us to help redesign its ATMs, we responded with curves. We introduced a subtle, curved cut-away Corian countertop that invites you to stand closer to the machine.

Strive to make the people who use your product or service feel welcomed and comfortable. At the Grand Wailea in Maui, you give your name to the guard at the front gate. A few hundred yards up the road, at the hotel entrance, you're then "magically" greeted by name. The warm welcome helps make a great first impression. It works partly because it's so personalized. One of the hallmarks of first-class service—on an airline or in any other environment—is that people address you by name.

2. MAKE METAPHORS

Try coming up with metaphors to inspire your new products. Several years back we were working on a portable disk drive called the Plus Passport. It had to be able to withstand being dropped from a couple of feet. You can express that in engineering terms, but the company had thought about this subject and managed to boil it down to a potent metaphor. "Think as though you're designing a thirty-five-millimeter camera," they told us. "Our drive should be just as rugged. Drop it six inches to the desktop, and you expect it to survive. Drop it three feet to concrete without its case, and you ought to blame yourself when it

breaks." It may sound like a little thing, but that image helped our team immeasurably in designing a product with just enough shock resistance.

Is there a guiding metaphor or phrase that might help you make your product or service focus on your customers' true needs and desires?

We all know how important names can be in defining and marketing products. Where would Apple be without the iMac, or Iomega without the Zip drive? That same spirit can be brought to themes or metaphors in the first stages of development. Encourage fun. Honda likes to come up with a philosophy, a mascot of sorts, to guide it through the design process. For a new engine it was designing several years ago, Honda came up with the concept of "a rugby player in a business suit," suggesting a balance between athletic prowess and gentlemanly professionalism. They even took it a step further with key phrases to capture specific attributes the car should evoke, everything from "tough spirit" to "stress-free." We've used metaphors on products from office chairs (the Vecta Kart chair that easily stacks, inspired by— you guessed it—a shopping cart) to soda cups (Pepsi's more durable, take-out Twist 'n Go cup is inspired by sports bottles).

Thinking up metaphors or phrases to guide your design and product objectives works even on small, internal projects. More and more companies have playful code names for products under development. Try it. You'll be surprised by how it can be a guiding light.

3. THINK BRIEFCASE

The traditional briefcase. It's a great way to help you think about products that bridge the gap between work and home. Why did the briefcase become such a powerful and loved icon for the American worker during most of the twentieth century? The classic American male worker loved his briefcase or lunch pail for the simple reason that it was one of the few things he could take home. You didn't lug your typewriter or desktop computer home on the train. People grew at-

The "stackable" Vecta Kart chair builds on the
metaphor of the shopping cart.

Steven Moeder

tached to their weathered leather briefcases the way they grow attached to an old wallet. Apple PowerBooks, Montblanc pens, and a handful of cell phones have all attained briefcase-style status. Judging from some of the projects we're currently working on under nondisclosure, I believe we're bound to see plenty more of these in the years ahead. Devices that cross over between work and home will always res-

onate stronger than purely office-based equipment. We might have the latest Intel processor at work, but we're emotionally attached to our PDA or cell phone.

Imagine that your ideal customer is on the commuter train, headed home after a long day. Make him or her want to bring your offering on that journey, and you can't help but succeed.

4. COLOR INSPIRES

Color works best when it's a pivotal, early step in design. I'm not an expert in color, but I've sat in on dozens of meetings with Mike Nuttall and listened to him make color part of the initial design goal. He'll ask a company what it wants to express with a product. If they say something along the lines of "powerful, for senior executives," Mike will come back with descriptions like "black, sleek, and angular." We made that machine for the Dynabook, and IBM has maintained that style—and color—to this day in its ThinkPad laptops.

OK, so you want your machine to be playful, nonconformist, express your artistic, right-brain tendencies. Well, you don't want black—or white. Maybe you're interested in lemon or orange. You've started on the path to an iBook, of course.

The key is starting early in the prototype stage. For all the color that's washed over consumer products in the past few years, most companies still live in a black-and-white world until the design and engineering are nearly complete.

Way before we had candy-colored Palms and iMacs, color made a huge difference. Olivetti introduced a rainbow of typewriters in the 1960s, equating color with, yes, portable typing—on the beach, in a sports car, and so on. In the flat plains of the Midwest, I've been able to spot from nearly a mile away whether someone is sitting on a green John Deere tractor or a red International Harvester. Color can be such a defining element in a product that certain companies seem to own pieces of the rainbow. Sony, for example, picked a distinctive orangish

yellow to boldly announce its durable "sports" product line; Tiffany, meanwhile, seems to have the retail franchise on a certain shade of blue.

5. BACKSTAGE PASS

Let your customers know what's going on behind the curtain, and they'll reward you with business and perhaps even loyalty. Early on, Amazon wisely started sending confirmation e-mails, letting customers know the progress of their orders through the shipping process. Netscape created animated shooting stars to let you know the browser was busily connecting; the hourglass icon on your screen lets you know that the computer hasn't forgotten about you. Philips introduced see-through kettles and toasters that let you see when the water is boiling and your toast has reached the right shade of brown. Clue people into what's going on behind the scenes, and they'll be happier, even if they're still waiting.

Help your customers poke their heads above the crowd. I bought insurance on-line recently, and they sent me mail or e-mail virtually every day to tell me where things stood, what the next steps were, and what they needed from me. Similarly, eBay—a Website I find practically addictive—sends you an update daily to remind you which bids you still have in play and which you've been outbid on. We all hate to be put on hold, but it's reassuring when the better "on hold" systems tell you how long you've got to wait before your call is answered.

6. ONE CLICK IS BETTER THAN TWO

Make your product or service work faster and simpler, and it will probably succeed. The one-click ordering process pioneered by Amazon makes a huge difference in the time-sensitive world of the Web (though it seems such a universal idea that I'm not sure it should be patentable). Off-line and on-line, there are lots of examples where companies make leaps by reducing clicks or eliminating steps. Have

you ever noticed that price never gets discussed in most FedEx shipments? Not only does it make businesspeople less cost-sensitive, but it also speeds up the processing of your package.

I have always liked the one-click feature on some ATM machines. After entering your password, you just press "Express $100," and you're on your way in seconds. The best Xerox copiers assume you want to make one copy on standard-sized paper—unless you tell them differently. Products need default modes. They should do the basics with minimal hand-holding.

What's the opposite of "one-click"? Unfortunately, I know an office phone system that fits into that category. When you call in for messages, the system asks your extension and password, then gives you *eight* different options. Unfortunately "listen to my messages" is not the first one, even though that's all you want to do 90 percent of the time. When you select "get messages" it still doesn't play them. Instead you are informed, in a painfully slow voice, the day, date, and exact time of the first message. The system then waits for further instructions, as if you haven't really been clear about what you want. Yes, you have to press *another* button to actually hear the call. And another to hear the next, and so on. No autopilot here. You've got to work to retrieve every single message.

Compare this to the Palm's simple operating system. Even when the product is off, you just push the phone-icon button, and it assumes you want to look up a phone number, since that's what most people do. Scribble the first letter of the last name of the person you're looking for, and up pops a screen of names, including probably the person you want to call.

What's the point? It's the twenty-first century. People are busy and often a little impatient. Don't let your cloud of features blur the simplest, most common use of your product.

7. GOOF-PROOF

Where would we be without the "autosave" and "undo" commands when writing a memo or letter? Nearly every product or service could use a "mistake-proof" button, a real or virtual "Wite-Out."

At IDEO we try to apply the "undo" principle to nearly every stage of product development. Take prototyping. Hot glue guns give us a margin of error. It takes about fifteen seconds for the glue to harden, plenty of time to wiggle the parts into position. All good products help prevent mistakes, but the real winners help you recover *after* you've misstepped: like self-sealing bicycle tires or the foldable side view mirrors on cars that spring back instead of breaking when you bump into them. How about your product or service? If your customer accidentally steps off the intended path, how easy is it for them to get back on track?

8. FIRST, DO NO HARM

The physician's creed applies to products and services as well. It sounds obvious, but too many companies forget this most basic principle. Take the pain or struggle out of your products or services, and you'll not only win over customers but likely beat out the competition as well.

Think of the traditional iron. There's something terribly wrong with a product that makes you spit at it or fry your skin to find out if it's ready. Why must crutches be so cumbersome and painful? Why must tin cans slice fingers? Who looks forward to lighting the pilot on a typical household water heater? (I always say my prayers first.) I'd argue that the reason this hasn't been redesigned is that manufacturers don't consider or care about the pain and fear homeowners feel.

Reducing danger or saving just a few lives can generate an awful lot of goodwill. Think of the benefits Volvo has reaped by making its cars safer.

Look around. Are you causing your customers discomfort or even minor pain in some small way? At my local grocery store, a blast of cold

air shoots down on your head when you enter. In rental cars, when you turn off the key in the ignition, the interior goes dark; couldn't a little LCD (they cost about eight cents) light up the door handle for a few seconds? And finally, as far as opportunities go, how about those hundreds of millions of CDs sold each year? I know retailers want to stop theft, but customers who just paid $16.99 for their new music shouldn't have to arm wrestle the package to get it open.

9. CHECKLIST

There's often a basic feature and compatibility list you need to check off before introducing a new product or service. Fail a critical compatibility test, for instance, and you'll flop. This is one element of product development that is entirely unforgiving. No matter how innovative your product may be, it must fit certain industry or public standards.

We once traded our services for stock in a computer company that flew in the face of this logic. In many ways, the Bytec Hyperion portable was superior to the first Compaq machine. The Bytec had a nicer screen, its keyboard and keypad neatly stowed away, and it was amazingly compact. But it was only 99 *percent* IBM compatible. That in itself wasn't remarkable, since even the Compaq lacked 100 percent compatibility. But the Bytec invited disaster by taking the unusual and ultimately fatal tact of advertising its incompatibility. In that 1 percent of incompatibility lay just enough free-floating anxiety to keep customers away.

Make a checklist of the "essentials" before you begin a project, the minimal elements your product or service needs in order to be accepted in the marketplace. Periodically check to make sure you haven't forgotten one of these basics.

10. GREAT EXTRAS

Great accessories or minor elements can carry a product, just as a great supporting cast can occasionally carry a movie whose star isn't up to par. Consider the features customers look for in a good cordless drill. A few years ago Panasonic had a cordless drill that sold extremely well not because it was the most advanced tool on the market, but because its *charger* worked better than most. It might seem like the tail wagging the dog, but the right accessory can make the product.

Automakers, genuine experts in the human psyche, understand this principle all too well. Think of what a difference little extras like comfortable seats make in your decision to buy a particular car. Or better drink holders. Or superior storage. Or even a better key.

People respond to the right small touches. It's up to you to come up with these surprises and fine details.

LIVE THE FUTURE

Rick English

Yogi Berra gave us the bad news: "Predictions are hard, especially when they are about the future." On the other hand, cyberspace novelist William Gibson gave us the good news: "The future has already arrived, it's just not widely distributed."

Knowing what's next isn't purely a matter of guesswork. Know enough futures and it's a little like putting together a puzzle. You've already got a lot of the right pieces, and the missing one could fall into place anytime. We're lucky at IDEO. We don't have to be wizards to know the future, at least the near-term future. Working today on tomorrow's consumer electronics, computers, medical devices, and office environments gives us an edge. You name it and I'll bet we heard about it a year or two before it hit the market. We started designing the first laptop for GRiD Systems in 1987. A few years later we were pushing the envelope of pen-based computing with GRiD, Go, and EO. This experience helped us enormously when Apple and Dell came to us to update their laptops and Palm and Handspring sought to redefine PDAs.

Around the time this book hits the stores you might also be able to read it as an electronic book, perhaps even one we helped design. Will you be talking to your appliances yet? I don't know for sure, but in the mid-nineties we demonstrated for Whirlpool how appliances could be operated with simple spoken commands. You could instruct your stove to roast a twelve-pound turkey by dinnertime.

Which brings me to a basic truth about the future. Not all futures happen at the same pace. Innovation, for example, tends to take root in the marketplace more slowly when there are deeply ingrained rituals, such as you find in the hearth and home. But many future products and services will adopt a more human, mobile face. For example, we developed a Q-shaped mobile work pod for Steelcase that you can literally drive into place. It has space for your computer, phone, and storage. Although it's on display at the Smithsonian's Cooper-Hewitt

Design Museum, I don't expect to see it in production anytime soon. We'll probably see mobile desks or work carts first.

Medical services and instruments are becoming more personal. The world of stethoscopes is just catching up with the Internet age. Our work in laptops and interaction design has greatly helped us move from cumbersome, fixed instruments to small devices that are portable and easy to use. We've helped develop simple, portable products to help patients test their glucose levels, their cholesterol, and their blood pressure. We've also designed remote devices that help patients with chronic illnesses keep their doctors apprised of daily changes in their symptoms. Watches, too, are becoming more personal, thanks to the latest round of innovation from Swatch. Tourists can now use a new technology embedded in the watch—Swatch Access—to buy tickets for ski resorts, museums, special events, and public transport. It's Dick Tracy for the tourist or commuter, and it's redefining the boundaries of an old tool. We've been doing similar work in the world of input and interface design. We've designed all kinds of input devices, from touch screens and trackballs to styluses and remote controls (for your entire house). Beyond our varied work in mobile and desktop computing, we've designed interfaces from tiny LCDs on medical instruments to interactive TV and entire Websites.

Seeing thousands of projects has afforded us a unique perspective on the future. We've gotten a sense of the holistic nature of change, learned how things evolve, and witnessed cross-pollination firsthand. Knowing the state of the art or the soon-to-be state of the art in one industry can give you a jump in your chosen field.

TOY WITH THE FUTURE

No one gets ahead by copying the status quo. The one thing Burton snowboards, e-Schwab, Yahoo, and Swatch all have in common is that they didn't imitate competitors.

Maybe your industry is different, but I can tell you that if you're in real estate, for example, you're not going to get many breakthrough solutions from fellow Realtors. So where should you start? Well, if you make products, you can't go wrong with your local toy store. Seriously. We've learned an enormous amount about how to do things simply and cheaply from toys, and we aren't the only ones. Lots of companies tear apart the latest toys to find out how the toy manufacturer did it with a few microchips and plastic. And we've spent more than a few afternoons watching folks play with the latest radio-controlled cars and planes.

Sure, you could do what people used to do: study advanced robotics or the latest research coming out of major universities. But we believe in the cheaper, faster, simpler approach. Choose the wrong metaphor, and you may go down the wrong path. You might peruse bicycle stores, for instance, to learn about new materials and shapes. Or find a passion for the high seas.

We've found sailing, with its wondrous mix of elaborate fittings and cables, to be inspiring. You never know where you might find something new. For example, the clutch mechanism that makes our Stella keyboard support magically stay in place at any height came directly from the sailing technique of wrapping a line around a nautical winch.

EMPOWERMENT PRODUCTS

Do you make products or services for the young? You probably couldn't lose by watching young skateboarders. If you sell products or services to older folks, retirement communities and coffee shops aren't a bad place to hang out. Getting older is about to become a very hot trend. Before long we'll have more time on our hands than ever before, and products that make seniors feel and act like they're thirty are bound to be in high demand. The perfect example is the electrically assisted bicycle that lets you take a long ride with only a minimal workout. Ad-

vances that give people feelings of empowerment without taking away the experience will be huge. Decades ago, we decided it was OK to let kids have more fun in baseball by allowing them to swat it out of the park with an aluminum bat. The future is in giving that option to everybody. Think about how sweet spots in golf clubs and lightweight, oversize tennis rackets opened up the games to novices. I couldn't do ten pull-ups if my life depended upon it, but the Gravitron machine at my health club lets me do a hundred—if I dial in the right body weight. What's next? Electronically assisted golf clubs and balls? The PGA would never approve them, but weekend golfers might suddenly be able to play scratch golf. Sure, purists will no doubt call it cheating, as they once complained snowboarding wasn't skiing, but how many of us wouldn't like to actually live out the fantasy of playing like Tiger Woods?

I also expect research and work for the disabled to start to spawn innovations for the general population. Just as curb ramps make it easier for everyone to carry rolling luggage, I believe work on eye- and voice-controlled interfaces for the disabled will soon reap more general benefits. You want to know what tomorrow holds? Seek out the early adopters and make camp at ground zero. Want to see the future of cellular telephony? Pack your long underwear and head for Finland, home to Nokia, perhaps the first place in the world where you can use a cell phone to buy a Pepsi from a vending machine.

Interested in electronic gadgetry? Take a trip to Akihabara, the blinking electronics hub of Tokyo. Want to know the future of interactive entertainment? Lunch in South Park, the epicenter of San Francisco's Multimedia Gulch. Where's Internet use headed? Just watch kids at school and the library. Looking for the future of athletic apparel? Head to the beach. Venice Beach, that is. Concerned about what new technical threats might upset your company's computer security? Hang out on Internet Relay Chat and listen to hackers brag about their latest exploits. Wondering what's next in venture capital? Breakfast at Buck's diner in Woodside, California, where the e-deals are struck.

Every industry has its hot spots, places where the *buzz* is, where you make connections and hear the latest. Yet it's not always so obvious. What if telecommunications is your thing? You might trek to the former chicken capital of Northern California, Petaluma, which seemingly overnight has become a new high-tech center (Cisco bought a little-known Petaluma company called Cerent for a cool $7.2 billion). Of course, the truly hot future spots don't exist yet, or didn't when I wrote this. You've got to find them yourself through the Web, searching out new magazines and doing old-fashioned human networking. Glean what you can from the world's top think tanks. But keep in mind that it's not the whole picture. Even an army of PhDs couldn't have told you snowboarding was going to be big, or that Hush Puppies were making a comeback.

Finally, cultivate and then listen to your own sages. You've probably got some very bright people right down the hall—if only you'd take time to get acquainted. Bill Moggridge, one of IDEO's original founders, is one of our biggest thinkers. Bill played a key role in many of the critical design shifts in computers. Back in 1984, the year of the Mac's introduction, he christened the birth of a new discipline at a major design conference. His first choice for a name was Soft Face, but a friend pointed out that it sounded a bit like a Cabbage Patch doll. So Bill coined the term "interaction design" to encompass the budding practice of applying industrial design skills to software interfaces. Of course, interaction designers didn't exist yet. No matter. Bill helped launch one of the world's best college programs in interaction design, and IDEO went on to hire many of these budding talents. Without that kind of vision and foresight, I doubt we would have stayed a step ahead.

MAKE CONCEPT CARS

Observing and listening are crucial. Finding "lead users" and quirky enthusiasts is even more essential. What else can you do to dance with the future? Why not try some concept projects on for size?

Automakers, of course, make concept cars primarily to promote their brands and let the public know that they're cool. We do concept projects for our own staff. Not just to give them a break from the daily routine but to help us come up with better ideas.

However creative we might like to think we are, we're all generally operating under a bundle of constraints. NASA scientists are extremely limited by weight, and Dell engineers worry about cost; I'm sure you've got a few things to worry about at your company too. Great concept projects are a chance to momentarily let all of that go—to permit people to exercise their creative muscles without the normal constraints.

I can understand if this sounds a bit radical. In my first years at IDEO I didn't quite believe in concept projects. I thought they were something other firms did because they couldn't land enough paying work. Then an IDEO team literally flipped on the light switch. Our light switch concept project had no boundaries—and it showed. Ten IDEO designers dreamed up fanciful light switches fashioned out of black rubber, pink acrylic with tassels, ABS plastic, and upholstery foam. A golden, ribbed fish diving into a rippled acrylic pond made the cover of *ID* magazine. The switches won design awards and were prominently featured in a show at New York's Museum of Modern Art.

Those vibrant switches changed my mind about concept projects. Not just because of the critical recognition Tim Brown and the rest of his San Francisco office received for their outlandish designs, but because the buzz that project sent through IDEO was immeasurable. The sense of possibilities and pride it gave our designers was astonishing.

Don't try to engineer a benefit in at the beginning of a concept project, or expect it to magically morph into next year's model. Whimsy is

critical. We did a concept project called "Keep on Flippin'" to reinvent the popular flip-flop beach sandal. One of our concept sandals looked like a wire whisk and made a very springy version of the requisite *flip-flop* sound. Others resembled mini-surfboards or evoked sixties furniture. Still another concept project, "the chocolate exploration," yielded products that were not only great fun but also delicious—like the nut-and-bolt-shaped chocolate confection.

Because people continually ask us for innovation, we're constantly thinking about the future. *Esquire,* for example, unexpectedly voted David one of the "21 Most Important People of the 21st century," along with such luminaries as Paul MacCready, Linus Torvalds, and Jeff Bezos. As part of the cover story, *Esquire* asked us to share one of our far-out ideas for the future. We sketched up a smart car, a podlike mobile work/live space that can link with others in "smart" highways. "It's all probably a ways off," *Esquire* wrote, introducing our sketches. "But someone's got to dream it before anyone can do it, no?"

True enough. In 1998 the BBC asked our London office to explore the future of the digital radio. This was by no means a typical concept project. The BBC had been making digital audio broadcasts since 1995, but high prices made for few manufacturers. So we came up with nine concepts for digital radios, including a splash-proof one. We also showed how you might display sports scores or financial information while playing a radio program. Or get more information about artists, recordings, books, or products. At the time, they seemed like pretty outlandish ideas. But after the BBC gave presentations on our digital radio concepts around the world, a European telecommunications company actually decided to make one.

So sometimes concept projects can lead directly to an innovation. But again, that shouldn't be your primary goal. They can also help you come up with novel metaphors or new architectures for product categories. Do a few concept projects and—like brainstormers—the exercise will become second nature. The next time you're struggling with a tricky project, close your eyes and pretend the barriers—size, materials, weight, or simple preconceptions—don't exist.

MAKE MOVIE TRAILERS

Concept projects help companies uncover clues to more distant futures. But companies more often are worrying about the near-term future, things they can build and services they can roll out within a year. Once you decide on what's possible, it's time to pull out the camera and start filming. Not a whole movie, because that would be expensive and you don't know the story yet. Instead, start with just a movie trailer.

You've seen hundreds of trailers in theaters. Everybody loves them. Little one-to-two-minute teasers. It's a great way to capture the essence of a project, the heart of the emotional experience you're seeking in a product or service. You don't have to figure out all the technical or marketing problems. The goal is a visual prototype. It can be as short as thirty seconds. We've filmed them for major international clients, and Skyline, our toy division, has made them in an afternoon to demonstrate a toy idea.

Try writing the print advertisement as you're developing your product or service. This forces you to find your unique selling proposition. The movie trailer and ads focus your team in a way no design or marketing plan ever could. And, of course, there's another advantage to making the trailer before you make the real thing. If you can't make a compelling trailer, maybe it's better to go back to the drawing board. Use Mickey Drexler's "T-shirt test"—if you can't make a T-shirt about it, maybe you don't have a compelling story.

Moviemaking supports our belief that products today are becoming more about experiences and that services are encouraging relationships. After looking at the future of computers, health care, offices, cars, and travel, we're more certain than ever that people are looking for rewarding journeys. Companies that can forecast little futures in shorts or trailers will have a great edge. They will be the ones to get their movies made.

READ ALL ABOUT IT

Lots of futurists make a very handsome living tracking trends and emerging technologies. Unfortunately, that's the *easy* part of the future game. The tough job is knowing which futures are going to come true.

You don't have to peer very far back in the past to see that plenty of things didn't happen quite the way we expected. During my childhood, most views of the future included personal helicopters, and sometimes even personal spaceships. Remember when the supersonic transport was going to revolutionize air travel and cut most flight times in half? Or when wind power was expected to eliminate our reliance on oil? The electric car was supposed to render the internal combustion engine obsolete. Even the pros strike out sometimes; in 1997 *Wired Magazine*'s cover story shoved a giant hand in our collective faces and shouted that "push" technology was about to render the Internet a passive medium. Oops. Not that IDEO bats a thousand either. We bet wrong on the acceptance curve for pen-based computing a decade before PDAs hit it big, and watched more than a few clients go bust trying.

What's a fun way to get a broad sense of the distant future? Science fiction isn't a bad place to start. If you'd read H. G. Wells at the turn of the century, you would have known that we'd put a man on the moon—just one of Wells's amazing predictions. Think of how our worldviews have been influenced by great works such as George Orwell's *1984* and Aldous Huxley's *Brave New World*. Arthur C. Clarke, who wrote *2001: A Space Odyssey*, predicted communication satellites decades before they were invented, and neatly framed the dilemma we face today in making our machines ever smarter without threatening our humanity.

We can thank William Gibson, author of *Neuromancer*, for a peek into a dark future in which biotech and cyberspace may make body parts and fantasies interchangeable (though some would argue this world already exists in Hollywood). Neal Stephenson, author of *Snow Crash* and *The Diamond Age*, seems to be one of today's hottest sci-fi

writers, spinning out visions of future societies. The designers of PDAs and mobile phones owe a debt of gratitude to the wonderfully elegant palm-sized phasers and radios featured on *Star Trek*. By the time you read this, there will be new sci-fi visions and bold predictions in the latest books and movies. You'll have to keep an eye out for them yourself in bookstores, theaters, and on the Web. And you'll have to separate the pulp from the prophets on your own.

PROJECT 2010

Futurists, sci-fi authors, and screenwriters can paint with a broad brushstroke. If in thirty or fifty years, just a handful of the developments they anticipated come true, they'll be considered visionaries, no matter how many things they got wrong.

We decided to take a leap of faith and do something crazy: predict the future for a time close enough that people might actually remember our prognostications. We already know the future two years out; that would be too easy. In the year 2000, we decided to take a stab at what some high-tech products might look like in 2010. It's not something we've ever done before on this scale.

And there was an added pressure. We decided to share our discoveries with *BusinessWeek,* which ultimately dedicated eight pages of its magazine to our findings. These are futures that we think are real. So how did we start thinking about the world of technology in 2010? A team of IDEO designers and engineers did a technology forecast and began collecting and synthesizing an extraordinary number of ideas. It helped that we've already done some futuristic work, like a recent collaboration with the MIT Media Lab on wearable computing. Less compelling futures quickly fell to the wayside. We started sketching, creating characters that might live in our future, imaging scenarios they might face.

Beverly Harper

The Kiss Communicator helps humanize
wireless technologies.

My favorite concept from Project 2010 is the "agent," the personal
digital assistant of the future. It's a combination Palm and electronic
wallet. It's your mobile computer/communicator, keys, and money, all
in one package. It's got a sleek metallic skin with a small porthole that
lets you see at a glance if you've received messages. Press your thumb-
print in the center to gain access to your data or to open car doors,
homes, or offices. Electro- and thermochromatic technologies will make
the agent's shell chameleon-like, reflecting your personality or mood.

We think products with personalities are here to stay. Even before
Project 2010, we dreamed up a concept of paired objects for lovers
called the Kiss Communicator.

You keep one of these sensuously-curved translucent objects for
yourself and give the other to that special someone in your life. Then
when you want to let them know you're thinking of them, you blow
into your own Communicator. It flashes briefly to let you know it's re-

ceived the gesture, and then transmits a signal to your distant lover. On the receiving end, only when they squeeze their Communicator with both hands, does it radiate a warm, rosy glow.

No one has picked up on the idea yet, but I think it's a winner. I'm sure teenagers would wear out the batteries in a couple of days, and high schools would quickly ban them, fueling the fire of interest.

Today people covet their PDAs and cell phones in a way they seldom have computers. I think we can expect to see small, personal devices become more beautiful, individualized, and reflective of our humanity. Think about it. How many times have you wished during a hectic day that you could blow a kiss to a distant special someone?

Our Tube concept is a natural evolution for IDEO, considering our work on every imaginable computer, from clunky desktops to tiny Palms. A metallic poster-shaped cylinder rolls out one or two flexible liquid displays. There is no CPU, keyboard, or mouse; voice recognition and pen-based computing will make keyboards superfluous. The Tube can open like a book for viewing text and graphics or be opened into a larger screen to watch video or surf the Web. We could use a new visual metaphor for a mobile, wireless world.

Then there's needed refuge offered by our Comfort Zone glasses. Engulfed by the cross talk of cell phones and mobile surfers, we're all going to need to momentarily escape—at least virtually. Comfort Zone glasses will be visual Walkmans. Surrounding noise will be electronically muted, and the eyewear will display a private screening of anything from data to films, all in the midst of the most crowded cybercafé.

We created walls and furniture in our offices that seem alive, with partitions doubling as displays and communication technologies. Holographic videoconferences will bring distant colleagues into meetings. At home we'll measure our vital signs daily with a Medical Mirror. Cameras and sensors will analyze everything from skin color to blood pressure and pulse.

If some of these ideas sound a bit far-out, consider that during one

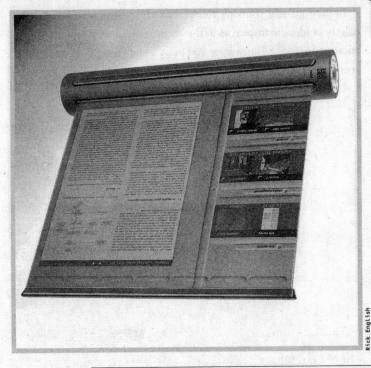

The "Tube" represents one possible scenario
for the future for reading.

IDEO brainstorm someone joked about programmable tattoos. Rather than discarding the wacky idea, instead we reeled it in. We brought the tattoo down to something possible in the year 2010—a souped-up runner's watch. A little translucent disk that sticks to your skin and picks up everything from your pulse to your pace. You can even transmit data to your agent to track your progress.

These are just a few of our future visions. We know they won't all come true, and even those that do may not arrive on schedule. But Project 2010 helped focus our thinking about the future. So often we're only thinking about the future we're interested in for today's project or client: the logical extension of a product, or next year's cool item.

Letting go and thinking ten years out gave us a wonderful new palette of ideas to inspire us in the years ahead. It also happened to be enormously fun. I'd encourage you to try a similar exploration on whatever scale fits your goals. Even the ones you haven't yet imagined.

Go out and make some futures of your own. Today.

GETTING IN THE SWING

ne aspect of "putting it all together" is knowing when to interrupt the process to "go live" with the current version of your new creation. We reach that point in every project. Sometimes multiple times. It's the time to cut and run. You can only brainstorm and prototype for so long.

Here's a simple drill that can help you to make the right decision—not just deep into a project, but right at the outset. Try turning your most feared deadline into a trusted friend. We try to work the deadline like 49ers quarterback Joe Montana used to work the two-minute clock. When we're at our best, that's how we approach the whole game. Nearly every product can be made incrementally better, including the one you've just finished. As we often say, that's why God invented clients—and bosses. Clients and bosses have budgets and schedules that bind the prototyping process. They're like the schoolteacher in the movie *Six Degrees of Separation* who says that the secret to getting great art from second graders is knowing when to take the paper away from them.

Success at innovation is like putting together the perfect golf swing. Back when I still attempted to play that challenging game, a golf instructor told me that there were just seventeen things to get right during the swing. Each one is pathetically easy: Can I keep my head down? Absolutely. Left arm straight? Can do. Knees slightly bent? C'mon, you're insulting my intelligence. They're *all* simple. The hard part is putting all seventeen together in real time (a lot harder than, say, patting your head and rubbing your stomach at the same time) and then doing it consistently several dozen times in a row out on the course. In my case, *many* dozen times.

Some of what we advocate is obvious, simple, and almost free. Stencil the brainstorming rules on your conference room walls and call a bunch of brainstormers to inaugurate the newly invigorated space. Buy

a few cases of giant Post-it easel pads and use them as freely as if they were little Post-its.

Tear up your "casual Friday" policy and adopt an "anything goes" approach, reminding people to use their good judgment when they know clients will be around. Double your budget for printed T-shirts. Or sweatshirts. Or messenger bags. If it's currently zero, of course, you have to more than double it, but you know what I mean. And then don't look too hard at what gets printed on the team-identifying apparel.

Try some homemade end-of-project awards, as a peer-oriented way to celebrate and recognize team contributions.

Publicly acknowledge a risk taker, a rule breaker, even a failure, and explain why every successful organization needs them. Hold an open house—internal at first—for staff members and project leaders to show off their work in process, as a way to motivate teams and encourage cross-pollination.

GIVE AND TAKE

Many aspects of reenergizing the culture are more subtle and built into tacit work practices. Here's an example of what I mean:

Your boss (or your client) gives you a month to come up with an important "deliverable," a piece of software, a report, a presentation, a product, or an ad campaign. We believe there are two dramatically different approaches to such an assignment. The first is to spend your month making the "perfect" version of the deliverable, polishing until it shines. Then, at the end of the month, you have the meeting with the Big Boss in which you—literally or figuratively—pull off the black velvet cloth and say *"voilà."* Well, if your boss throws up all over the thing, you're in trouble. Ego damage, for sure. Maybe even status and career damage, depending on your boss.

The second approach to that same one-month challenge is to burn the whole first week cranking out four or five really crude outlines or prototypes. The high-tech one. The playful one. The low-cost model. The pure-digital version. Then you squeeze in a ten-minute meeting at the end of the week with the Big Boss. Even in the unlikely possibility that she hates *all five* of your ideas, you're going to learn a lot as she tells you what's wrong with them, and you've now got three weeks to make the sixth one really sing.

Chances are, she'll pick elements from two or more of the prototypes, and you'll be able to combine the best of each in your final version. Even so, whatever criticism you get in week one doesn't sting much. After all, it's not your finished work, and you haven't put too much ego (or career risk) into any of the alternatives.

If you take this message to heart, you'll have to start training your boss, getting him or her accustomed to the idea that you'll be back around, long before the deadline, to get solid feedback. Try an up-front deadline compromise like "May 30 is great, if you'll let me have ten minutes with you on May 7 to make sure I'm on course."

As long as you negotiate it up front, it's such a reasonable request that who could say no? And if *you* are the manager or executive we're talking about, even better. All you have to do is hold your tongue and keep your criticism focused on the objective. What I mean is developing an attitude in which everyone around you understands it's OK to show you rough prototypes. No making fun of the workmanship or materials. No fussing over typos, no critiquing the fit and finish of a prototype. *It's just a prototype, after all.* Learn to restrain from issuing these little barbs, and you'll see more good work sooner, get a chance to redirect projects headed off into the weeds, and end up with better final results.

HITTING THE PRACTICE RANGE

Over the years we've come up with some valuable innovation practice tips. Try jotting these down in your own words and posting them around your workplace. Most of all, practice them whenever you can.

- Watch customers—and noncustomers—especially enthusiasts.
- Play with your physical workplace in a way that sends positive "body language" to employees and visitors.
- Think "verbs," not "nouns," in your product and service offerings so that you create wonderful experiences for everyone who comes into contact with your company or brand.
- Break rules and "fail forward" so that change is part of the culture, and little setbacks are expected.
- Stay human, scaling your organizational environment so that there's room for hot groups to emerge and thrive.
- Build bridges from one department to another, from your company to your prospective customers, and ultimately from the present to the future.

ADOPTING AND ADAPTING

It's time to get out there and give it a whack.

Building a culture of innovation is as simple—and as hard—as sifting through the ideas in this book and adopting (or adapting) the ones that fit your situation. Most of IDEO's examples come from the tangible world of products, but we have already proven to our clients that we can adapt those same approaches to everything from shopping for

your first home to on-line banking or taking a ride on a high-speed train.

Try it yourself. Innovation isn't about perfection. You've got to shank a few before your swing smooths out. Get out there and observe the market, your customers and products. Brainstorm like crazy and prototype in bursts.

You know the drill. The next time you're knee deep in a challenging project, don't forget the true spirit of innovation.

That's right. Have some serious fun.

Tom Kelley
tomkelley@ideo.com

Abyss, The (film), 93, 104

Adoption curves, 167, 173–74, *173*, 285

Advanced Cardiovascular Systems Indeflator, 25–26

Aerobie football, 110–11, *111*

Aerogel, 144

Affordances (handshakes), 165, 182–85; bad, 183; homespun, 184–85; lack of, 183; new, 184; opportunity and, 184–85

Alcoa, 138–39

Amazon.com, 97, 107–8, 134–35, 231, 270, 271

American Corporate Hall of Fame, 181

Amtrak Acela designs, 129–30, 195

Anderson, Carl, 12

Answering machines, 175

Apple: Duo Dock, 114–17, *116*; evangelism of, 189–90; iBook, 185, 189, 259, 269; iMac, 76, 114, 145, 183, 212, 231, 259, 267, 269–70; Lisa, 19, 189; Macintosh, 70, 130, 189–90; monitor stand, 233; mouse, 4, 19, *20*, *101*, 113, 189, 256, 264–65; Newton, 172; PowerBook, 158, 189, 269, 277; prankster environment, 18

Art of Possibility (Zander), 88

Arthur Andersen, 55

Ask Jeeves, 188

Atari, 97

Atlas Snowshoe Company, 30

AT&T, 40–41, 89, 137, 191; Allegro design language, 259–60

Automobiles: affordances in, 183–84; concept cars, 282–83; drink holders in, 167–68

Bank of America, 36

Barriers and obstacles to innovation: adoption curves, 167, 173–74, *173*; affor-dances (handshakes), 165, 182–85; company's mind-set, 165, 172–73; creativity checklist, 180–81; cultural resistance, 165, 166–71; customer perception/product use, 165; deal killers, 176–77; effort and persistence to overcome, 191–92; evangelism (promotion) to overcome, 165, 189–91; FUD factor, 176; metric vs. imperial system, 169–71, *170*; over-coming or side-stepping, 165, 171–76, 189–92; PalmPilot, 165–66, 171–74, 183; patents, 177–78; rituals and traditions, 165, 185–89; skill sets, 181–82; thresholds of accep-tance and rejection, 165, 175–76

Bass, Ron, 86–87

Baxter Healthcare, left ventricular assist system, 36–37

Bayer aspirin, 186

BBC, 283

Beach chair, Sun Tracker, 155, *156*

Becker, Franklin, 131–32

"Beginner's mind," 60

"Being left-handed" principle, 33–34, 36

Bell & Howell, 199–200

Bellagio, 205–6

Ben & Jerry's Ice Cream, 75

Berkley Outdoor Products, 51–52, *52*

Berra, Yogi, 277

Bezos, Jeff, 107–8, 134–35, 283

Bezos, MacKenzie, 108

Billing, Tim, 161

Biometrics access device, 175–76

Blakely, Dave, 83, 191–92

"Bodystorming," 62

Bon Appetit, 207

Boyle, Brendan, 110–11, *111*, 246–47

Boyle, Dennis, 17, 18, 77–81, 94, 106, 113, 115, 131, 158, 161, 261, 262, 264; Tech Box and, 142, *143*, 144–46

Bradley, Paul, 103, 265

Brainstorming, 5, 53, 55–66, 246–47; "bodystorming," 62; boss at sessions, 65; broad spectrum of participants, 65; build and jump, 58–59; content-related homework, 60; customer needs over organizational goals, 57; effect of, 62–64; example, wine containers, 60–61, *61*; frequency needed, 55–56; get physical, 61–62; group warm-ups, 59–60; ideas per hour, 58; integrating into organizational culture, 55; length of session, 56; materials and products brought to, 62; no note-taking, 56, 66; no turn-taking, 56, 65; number ideas, 58; on-site vs. off, 56, 65; rules, playful, 57–58, 66; seven secrets for better, 56–62; shopping cart, 10; six ways to kill, 64–66; spatial memory and low-tech tools, 59; topic statements, 56–57

Brave New World (Huxley), 285
Breathe Right nasal strips, 156–57
Britain: cultural resistance to new products, 168; metric system adoption, 171
Brown, Tim, 282
Buck's Diner, Woodside, CA, 280
Burton, Jake and Burton Snowboards, 176, 238; goggles for, 238
BusinessWeek /Architectural Record Award, 139; IDEO predictions, 2010, 286–90
Bytec Hyperion, 273

Caesar's Forum, Las Vegas, 206
Cameron, James, 104
Cassidy, John, 235
Cattle guards, 169
CDs, 188–89
Cell phones, 166–68, 175
Central Market, 206–7
Cerent, 281
Chance. *See* Unpredictability
Charles Schwab brokers, 232–33; e-Schwab, 249
Chevys restaurants, 208
Chiat/Day, 134

Christianson, Tom, 177
Cisco Systems, 281
Clubhouses, 91–92
Cocktail napkins, 181
Coinstar, 190–91
Concept projects, 282–83
Cook, Scott, 30–31
Corcorran, Sean, 32–33, 45–46, 49–50, 106, 186–87
Craig, Roger, 222, 224, 225
Crick, Francis, 109
Cross-pollination, 49–50, 149, 156, 158–62, 278
Customers: arrogance toward, 41; brainstorming and, 57; empathy and, 41–42; making them feel like heroes, 51–52

Dancall cordless office phone, 46–47
Davies, Scott, 222, 224, 225
Datum shelf, 32–33
Dayton, Douglas, 17, 18, 126–28, 200
Dell computers, 212, 277
Diamond Age, The (Stephenson), 286
Digital radios, 283
Disney Stores, 201, 203
Drexler, Mickey, 117–18; T-shirt test, 284
Dubinsky, Donna, 78–79
Dymo label makers, 178
Dynabook laptop, 62–63, *63*, 151, 269

eBay.com, 245, 270
eBooks, 189
Edison, Thomas, 69–70, 87
EFX, elliptical exercise machines, 28, 29
Elahian, Kamran, 151
E-commerce, 176, 231–32, 245
E-mail, 184, 188; surveillance, 84, 175
"End of the Job, The," 121
Enorme phone, 233, *234*, 235
Entrepreneurs, 166, 176–77, 190–91; hot groups and, 100; observing and, 29–31
Epson, 184
Ericsson, 123–24, 132

Esquire, 283
Experience Economy, The (Pine & Gilmore), 196
Experiences, creating retail or service: airport experience, *193*, 196–99; Apple computers, 212; applying to all businesses, 211–12; Bell & Howell, 199–200; Bon Appetit, 207; Central Market, 206–7; Chevys restaurants, 208; child's birthday, 196; customer journey, following, 199; Dell computers, 212; designed experience, 10 steps, 196–99; GE sales incentives, 196; Hank Haney Golf complexes, 207; hotel industry, 214–15; IDEO's front lobby, 209; IDEO's job applicant process, 216; innovation in small improvements, 200–201; JanSport, 215; Kinko's, 209–10; learning from Vegas, 205–6; Lexus, 195, 212–13; Martha Stewart, 210; Medtronic, 200–201; Merloni, 200; negative, changing perception of, 215–16; Nordstrom, 195; Pike Place Market, 208, *209*; Restoration Hardware, 210–11; retailers, 201, *202*, 203; services, rethinking, 214–15; sign-in books, 216–17; Starbucks, 211; Steelcase, 203–4; tour badges, 217; travel, 208; Webvan, 195

Failure and success, 232–33, *234*, 235, 296
FedEx, 44, 271
Feuhrer, Jim, 11, 128
Field trips, 94
"Fine" response, 27
Finger Blasters, 94
Finland, 168, 280
Fishing kits, all-in-one, 25, 51–52, *52*
Footrest, adjustable, 36
Free Willy (film), 93
Frisbee, 259
Fry, Art, 261
FUD factor, 176
Fukasawa, Naoto, 262–63
Future, innovation for the: concept projects, 282–83; empowerment products, 279–81; IDEO predictions, 2010, 286–90; movie trailers, 284; predictions, false or accurate, 285–86; products in near, 277–78; toys as predictors, 278–79

General Electric (GE) Mark Makers, 196
General Motors (GM), 41; EVI electric car charger, 122, 176
Gibson, William, 277
Gillette: Liquid Paper, 244; Mach 3, 242–43
Gilmore, James H., 196
Go Corporation, 172
Golden Gate National Parks, 117–18
Golf swing, *291*, 293
Goodby, Jeff, 118
Gordon, Stephen, 210–11
Gore, Al, 191–92
Graffiti, 166, 172–73
Graves, Michael, 249
GRiD Systems, 152, 166, 171–72, 277

Hamel, Gary, 3
Han, Amy, 224, 262, 264
Handspring, 136, 264, 277; Visor, 46, 77–81, 89, *170*, 171, 183, 184
Hank Haney Golf complexes, 207
Haring, Keith, 70
Harley-Davidson, 188
Harmon, Butch, 5
Hawkins, Jeff, 78, 79, 165–66, 172–73, 261
Haygood, David, 94, 136–37
Heartstream, 213, *253*, 258–59
Heat pipe, copper, 144, 158
Hertz NeverLost, 257
Hewlett-Packard, 134, 140, 172
Hexcel, 94
Hiring employees, 121; IDEO, 88–89, 214; outsiders, 161
Holtz, Lou, 84–85
Homebrew Computer Club, 92
Hotel industry, 214–15

Hotmail, 231
Hovey, Dean, 17

IBM ThinkPad, 269
Idea wading, 159–60
IDEO: approach to innovation, 5; biking at, 47, 94, 128, 160, 238; brainstorming at, 55–56, 60–61, *61*, 66, 155, 246–47; buying the products they designed, 89; clubhouses at, 91–92; concept project, chocolate exploration, 283; concept project, "Keep on Flippin'," 283; concept project, light switch, 282; "crazy" characters at, eight categories, 98–100; cross-pollination approach, 49–50, 149, 156, 158–62; culture of creativity at, 18–21; description of, 4–5; end-of-project awards, *67*, 89, *90*, 91; end-of-year bashes, 93; fail to succeed philosophy, 232, 233, *234*, 235; field trips at, 94; film projects, 93, 104; growth of, 3–4, 14, 72; hiring process at, 88–89, 214; hot groups at, 70–73, 75, 76–82; Industrial Design Excellence Awards, 4; launch of, 17–21; library, 139–40; lobby and entrance displays, 140–41, 209; lunches and team meetings, 95–96; machine and model shops, 99, 141; magazine subscriptions at, 159–60; methodology, 5-steps, 6–7, *see also,* Brainstorming, Launching a project; Prototyping; morale, 141; multidisciplinary approach, 39; *Nightline* program, *1,* 8–14, 70–71, 76, *90,* 91, 155, 213; office, Boston, 126–28, *127*; office, first, 123; office, Grand Rapids, 77; office, London, 283; office, Munich, 243; office, San Francisco, 139–40, 282; office size, 123, 136; office, Tokyo, 262–63 office under the wing, *15,* 95, 130–31; offices, High Street, 122–24, 128–29; pranks at, 94, 125; predictions, technological, for 2010, 286–90; product failures, 151–52, 285; products and projects by, *1,* 4, 7–8, 19, *20,* 25, 32–37, 45–52, 62–63, *63,* 75, 76–81, 79, 89, *101,* 103, 106, 113, 114–17, *116,* 122, 127–28, 129–30, 154–59, *156, 157,* 174, 175–76, 178, 185, 186–87, 189, 195, 199–201, 203–4, 231–36, *234,* 238, 246–47, 255, 256, 258–65, *263,* 266–74, *268,* 275, 277, 279, 282–83, 286–90, *287,* 288; rules, minimal, 243; S-curve and, 174; Sand Hill Challenge, *219,* 221–26, *227,* 228; Skyline, 246–47; Spunks, 130–31; status at, 9; T-shirts and other solidarity symbols, 96–97; team icons, 130–31; Tech Box, 142, *143,* 144–46, 158–59; tour badge, 217; tours at, 92, 141; transition, as David Kelley Design, 21; transition, merger with other companies and name IDEO created, 21; "Under the Wing" lunches, 95; University (IDEO U), 92–93; work space, *119,* 123, 124–25, 126–30, *127*
ID Two, 21
Igloo, 247–48
Industrial Light and Magic, 75, 94
Innovation: begins with an eye (observation), 26–52; as centerpiece of corporate strategies, 3; creativity list, 180–81; culture to support creativity, 13–14; decision to implement, 293–97; importance of, 13; integrating into organizational culture, 294–95; opening the floodgates, 179–81; pace of, 277–78; practice tips, 296; process of, 13–14; small crashes, 226. *See also* Barriers and obstacles to innovation; Brainstorming; Launching a project; Prototyping; Refining products; Risk Taking; Teams (Hot Groups); Workplace
Innovation Network, 251
Inspiration: adversity and, 128–29; by observation, 36–37; sources, 278–79; tech box, 142, *143,* 144–46, 158–59
International Harvester, 270

Internet: auctions, 245–46; car buying on, 50; e-commerce, 176, 231–32, 245; future developments, 280; open source software, 244–45
Intuit, 30
Iomega, 267
Iridium, 41
Isuzu, 191
iWon.com, 231

JanSport, 215
Japan: Alihabara, 280; cultural resistance to American products, 167, 168; steam irons, 255
Jehl, Francis, 70
Jobs, Steve, 19, 75–76, 130, 189, 259
John Deere, 270
Juggling for the Complete Klutz (Cassidy), 235

Kelleher, Herb, 240–41
Kelley, David, 142; alma mater, 223; AT&T account and, 191; at Boeing, 17, 85, 162; childhood "prototyping," 105; coloring outside the lines, 246; drawing people and, 182; early jobs of, 17; *Esquire* article, "21 Most Important People of the 21st Century," 283; first hire by, 98; launch of IDEO, 17–21; at NCR, 162; office of, 126; PalmPilot and, 165–66; rules and, 243; shopping cart, *Nightline* program, 11; Sottsass, Ettore, and, 233, 235; Stanford University and, 259; team creation at IDEO, 72–73
Kelley, Tom, 4, 9, 121; AT&T account and, 191; consulting firm interview, 132–33; GE Mark Makers and, 196; hot group experience, Singapore, 81–82; as management consultant, 240; office space measurement, early job, 137; Palm V and, 174; Shoe Care project, 187
Kennedy, John F., 75

Kettering, Charles F., 41
Kevlar, 144, 153–54
Kimberly-Clark, Kleenex, 153
Kinko's, 209–10
Kiss Communicator, 287–88, 287
KitchenAid, 169
Klebahn, Perry, 29–30
Kleiner Perkins, 151
Kodak, 42
Kurjan, Chris, 145

Las Vegas, design experiences, 205–6
Launching a project, 7, 25–52; adjustable footrest, 36; AT&T business strategy, 40–41; balloon angioplasty device, 25–26; Baxter Healthcare system, 36–37; "being left-handed," 33–34, 36; bicycle water bottle, 25, 47–49, 48; bug list, 28–31, 43–44; cross-pollination approach, 49–50, 156, 158–62; customers, empathizing with, 41–42; Dancall cordless office phone, 46–47; Datum shelf, 32–33; elliptical exercise machines, 28, 29; embracing your crazy user, 37–39; finding rule breakers, 39–41; fishing kit, 51–52, 52; human factors, and "little" innovations, 44–49; "innovation begins with the eye," 28; inspiration by observation, 36–37; kids, talking and listening to, 33–34; Kodak digital imaging, 42; looking cross-eyed, 154–58; medical mobile computer design, 45–46, 49–50; NEC's Versa products, 38–39; no dumb questions, 31–33; observation exercises, 42–43; observing people vs. asking, 26–28; Philips, kids' mouse, 34; Procter & Gamble, Crest tube, 45, 89; sensory immersion, 31; shopping cart, 9–10, 155; toothbrush, Squish Grip, 34, 35; verb products, 46–49, 175, 261
Ledbetter, Carl, 255, 260
Lee, Alex, 131
Lego, 188
Lexus, 195, 212–13

Liddle, David, 236
Linux, 74–75, 244–45
Liquid Paper, 243–44
Lockheed Skunkworks, 70
Logitech, 110
"Looking cross-eyed," 149, 154–58, *156*, *157*
Lucas, George, 94
Lyons, Dave, 221–22, 223

MacCready, Paul, 283
Mannock, Jerry, 19
Martha Stewart, 210
Matrix, 21
Matsushita, 255
McGaw Horizon infusion pump, 261
McKenzie, Gordon, 251
McKim, Bob, 17
Medical devices and systems, 7–8; defibril-
 lators, 200–201, 213, 253, 258–59;
 Indeflator, 25–26; infusion pump,
 261; mobile computer design, 45–46,
 49–50; pacemaker, 213–14; personal-
 ization of, 278; simplifying, 258–59;
 skin for burn victims, 122; ventricular
 assist system, 36–37
Medtronic, 200–201, 213–14
Merloni, 91; Home Smart Monitor, 200
Metaphor workstation, 236
Metcalfe, Bob, 27–28
Metric vs. imperial system, 169–71, *170*
Microsoft Word, 257
Miller, Larry, 28, 29
Mirage casino, Las Vegas, 214–15
Moggridge Associates, 21
Moggridge, Bill, 281
Mohr Davidow, 221, 225
Molbak, Jens, 190–91
Momenta, 151–52
Morale, 93–94, 141
Mouse: Apple, 4, 19, *20*, *101*, 113, 189,
 256, 264–65; Catcher, 183; cordless,
 235–36; iMac design problem, 183;
 Philips, for kids, 34
Movie trailers, 284
Multidisciplinary approach, 39
Mylar, 144

NASA, 75; Mars Climate Orbiter disaster,
 169–70
NEC's Versa products, 38–39, 199; Green-
 house, 39
Nesmith, Bette, 243–44
Neuromancer (Gibson), 285
Newman, Paul, 75
Nightline program, "The Deep Dive: One
 Company's Secret Weapon for Inno-
 vation," *1*, 8–14, 70–71, 76, *90*, 91,
 155, 213
Nike sunglasses, 89
NikeTown, 201, 203
1984 (Orwell), 285
Nitinol "memory metal," 144, 158
Nokia, 201, 280
Norand Pen*Key computer, 174
Nordstrom, 195
Nuttall, Mike, 185, 191, 265, 269

Observation, 28–52; "being left-handed"
 principle, 33–34, 36; exercises,
 42–43; inspiration by, 36–37; "why"
 questions and, 37. *See also* Launching
 a project
Omnicon Group, 134
O'Neill, Paul, 138–39
Orange telecommunications, 283
Oticon, 82
Otis Elevator, 127–28
OXO International, 131

Pack-A-Cino, 144
Palm Computing, 172, 277; Pilot, 165–66,
 171–74; Palm V, 4, 46, 76, 89, 189,
 261–64, *263*, 269, 271; Zoomer, 172,
 173
Panasonic, 274
Park Hyatt Hotel, Los Angeles, 215
Patents, 177–78
Pauling, Linus, 55, 58
Pepsi, 267
Personal Digital Assistants (PDAs),
 151–52, 165–66, 171–74, 183, 269,
 277, 285, 286, 288–89

Petaluma, CA, 281

Peters, Tom, 74, 160

Philips: mouse for kids, 33–34; see-through appliances, 270

Pike Place Market, 208, 209

Pine, Joseph, 196

Pixar, 125–26

Plus Passport, 159, 266

Polaroid I-Zone cameras, 89, 152–53

Post-it, 261

Precor, Inc., 28

Procter & Gamble: Crest Neat Squeeze tube, 45, 89; Ivory Soap, 150; Reflect.com, 242; risk taking and R&D, 242–43

Prototyping, 5, 7, 101, 103–18, 111, 116; Aerobie football, 110–11, 111; Amazon.com start-up, 107–8; Apple Duo Dock, 114–17, 116; bad ideas first, 114; Bell X-1 aircraft, 109–10; build to learn, 106; client meetings and, 106; DNA model, 109; film-making as, 104; Golden Gate National Parks, 117–18; incremental reviews, 112–13; innovation and new ideas from, 108–11, 111; mouse (Apple), 113; movie trailers (visual prototype), 284; new service or business, 107–8; 117–18; pictures, reports, or drawings vs., 111–12; playful approach to, 105; as problem-solving, 103; rule-bucking, 117–18; shopping cart, 1, 10–12, 155; snowboarder goggles, 103, 238; soap box derby racer, 222–25, 226; video game controller, 110; Wells Fargo ATM security, 113; workspace and, 129–30

Quicken, 31

Refining product design, 253, 255–67, 263, 268, 269–74; backstage pass, 270; checklist, 273–74; color inspires, 269–70; fighting feature creep, 255–57; first, do no harm, 272–73;

Frisbee, low mass solution, 259; goof-proof, 272; great extras, 274; make a great entrance, 266; make metaphors, 266–67; Mouse, redesigning, 264–65; one click is better than two, 271–72; Palm V, 261–64, 263; simplifying directions, 257–59; think briefcase, 267, 269; Wet Nap Interface, 255–56

Reflect.com, 242

REI, 75, 202, 203

Rental cars, 50–51

Restoration Hardware, 210–11

Rice, Jerry, 156

Risk taking (coloring outside the lines), 229, 231–33, 234, 235–36, 237, 238–49, 250, 251; being big, acting small, 241–43; breaking the rules, examples, 244–49, 250, 251, 296; Charles Schwab, 232–33; cordless mouse, 235–36; excessive, 251; fail to succeed, 232–33, 234, 235; Gillette, Mach 3, 242–43; iMac, 231; Linux, 244–45; Liquid Paper, 243–44; mountain biking, 238–40; on-line auctions, 245–46; Procter & Gamble, 242; Silicon Graphics, 231; snowboarding, 236, 237, 238; Southwest airlines, 240–41; Web ventures, 231–32, 249

Rubbermaid, 251

Saccharin, 150

Safe Rides, 221

Samsung, SyncMaster monitors, 152

San Jose Tech Museum, 94

Sand Hill Challenge, 219, 221–26, 227, 228

Saturn, 50

Schwab, Charles, 26

Schwab, Michael, 118

Schwinn, 238–39

Sculley, John, 116–17, 172

Segal, Leon, 28

Sephora, 248

Sharper Image, 177

Shelves, flexible office, 25

Shoebox Greetings, 248

Shopping cart, *1*, 8–14, 213; brainstorming, 10, 155; end-of-project award, *90*, 91; hot group and, 70–71, 76; investigating the project, 9–10; prototyping, 10–11

ShowerTek fog-free shaving mirror, 177

Shubert, Larry, 104

Silicon Graphics, 231

Silverstein, Rich, 118, 191

Singapore Airlines, 81–82

Six Degrees of Separation (film), 293

Skillman, Peter, 8–9, 131

Smith, Jack, 12

Smith Corona, 178

Smith Sports Optics: Roll-offs, 157–58, *157*; ski goggles, 89

Snowboarding, 236, 237, 238, 281

Snowcrash (Stephenson), 285–86

Snowshoes, 29–30

Sony, 270

Sorbothane, 159

Sottsass, Ettore, 233, *234*, 235

South Park, 280

Southwest Airlines, 240–41

Specialized Bicycle Components: mountain bikes, 239–40; Racer's Edge water bottle, 25, 47–49, *48*; slogan, "Innovate or die," 240

Speed (development and marketing time), *219*, 221–28

Stanford University, 259

Starbucks, 212

StarTac, 261

Star Trek, 286

Steelcase, 32, 145, 203–4, 277–78

Steele, Darrin and Dan, 225

Steele, Fritz, 131–32

Stella keyboard-support, 279

Sun, Rickson, 17, 161, 178

Sun Microsystems, Java, 258

Suri, Jane Fulton, 37–39, 161, 265

Swatch, 249, *250*, 278; Access, 278

Switky, Andy, 264

Target, 249

Teams (hot groups), *66*, 67–100; abolish "they," 83–84, 85–86; abundance mentality, 85–86; Bass, Ron, team approach of, 86–87; brainstorming and, 64; building a team, 88–89; challenging goals, 77–81; characteristics, 71; clear, tangible goal and serious deadline, 69, 75–76; close quarters and settings, 81–83; clubhouses and, 91–92; competition and playful rewards, 73–74, 89, *90*, 91; Eyemodule project, 77–81, *79*; icons, creating, 130–31; IDEO's Hollywood-like system, 71–72; individuals vs., 86–87; information sharing, 91–93; meetings (energizing ideas for), 95–96; members and eight crazy characters that can jump-start a group, 97–100; morale building, 93–94; motivation, passion and, 74–76; myth of lone genius, 69–70; opportunities in obstacles, 76–77; projects and products spawned by, 70; right mix, 83; Sand Hill Challenge, *219*, 221–26, 227, 228; setting the bar high, 75–76; Southwest Airlines, 241; T-shirts and other solidarity symbols, 96–97; trust and, 84–86, 175; turn work into "wow" projects, 74; variety or simultaneous projects, 87; workplace and, 123, 134–36

Tech Box, 142, *143*, 144–46, 158–59

Telephone design, 260

Telesensory Systems, 161

Texas Instruments, 256

Thalheimer, Richard, 177

Thompson Aviation Library, 130

3Com, 184; EtherLink card, 27–28

Tiffany, 270

Timing, 152–53, 211

Toothbrush, Oral B Squish Grip, 34, *35*, 260–61

Toothpaste, 25, 45, 89

Torvalds, Linus, 244–45, 283

Toys, ideas and technology source, 279

Travel experiences, 208

Tube, *288*, 289

2001: A Space Odyssey (Clark), 285

Union Carbide, 133

Unpredictability and chance, fostering innovation from, 149–62; balancing and famous examples, 150; change hats, 161–62; cross-pollination, 49–50, 149, 156, 158–62, 278; cross-train, 162; hire outsiders, 161; hold an open house, 160; idea wading, 159–60; inspire advocates, 161; "looking cross-eyed," 149, 154–58, *156, 157*; play director, 160; product adapted into other uses, 156–57; product failures, 151–52, *see also* Risk taking; quirky uses for products and, 153–54; timing and, 152–53, 211; trends, 154

U.S. Mint, coin introduction, 177

Vacuum cleaners, 167

VCR, 256, 248, 261

Vecta, 106; Kart chair, 267, 268

Velcro, 150

Venice Beach, CA, 280

Verb products and thinking, 46–49, 175, 261, 296; experiences, 195

Volkswagen Beetle, 70

Warner Brothers Studio stores, 201, 203

Watson, James, 109

Webvan, 195

Weinstein, Richard, 169

Wells Fargo, ATM machines, 113, 266

Whirlpool, 277

Wine cork, 186–87

Wired magazine, 285

Wolverine Worldwide, Hush Puppies, 154, 281

Woods, Tiger, 5

Workplace and working environment, 121–46; Alcoa, 138–39; Amazon.com, 134–35; AT&T, 137; blend of openness and privacy, 124; "body language" of, 131–33, 296; building blocks, 126–30, *127*; building neighborhoods, 122–24; Chiat/Day, 134; contests for design, 128; cost, 121; crowded conditions vs. excess space, 135–36; David's office, 126; displays of history and achievements in, 140–41; egalitarian culture and, 132, 134; Ericsson's, 123–24, 132; giving workers a view, 138–40; Hewlett-Packard, 134, 140; hierarchy and, 125, 126, 132, 133, 136–37; IDEO's, *119,* 122–23, 124–25, 126–30, *127,* 139–46; inspiration from adversity, 128–29; library, 139–40; open office space, 133–34; personal ownership of, 124–26, 129; Pixar's, 125–26; prototyping, 129–30; simple team space, 133–36; team icons, 130–31; Tech Box, 142, *143,* 144–46; Union Carbine, 133; Yahoo!, 135

Workplace by Design (Becker & Steele), 131–32

World Wide Web. *See* Internet

Wigglewriter, 63

Wynn, Steve, 205

Yahoo!, 97, 135, 188

Yeager, Chuck, 109–10

Yousefi, Andre, *90*, 91

Yurchenco, Jim, 17, 18, 113, 161, 264

Zander, Benjamin, 88

Zelco, 154

Zeller, Noel, 154

ABOUT THE AUTHORS

TOM KELLEY

Tom Kelley is General Manager of IDEO, the world's leading design consultancy specializing in turnkey product development and innovation strategy. Working together with his brother, IDEO founder and chairman David Kelley, he has helped manage the firm as it has grown from 20 designers to a staff of over 300. During that time, he has been responsible for diverse areas such as business development, marketing, human resources, and operations. Like everyone else at IDEO, he also occasionally gets down on his knees to cut foam core alongside IDEO clients and designers to support the firm's brainstorming and prototyping efforts.

JONATHAN LITTMAN

Jonathan Littman is the author of *The Fugitive Game*, *The Watchman*, and *The Beautiful Game*.